Event-Triggered Transmission Protocol in Robust Control Systems

Controlling uncertain networked control system (NCS) with limited communication among subcomponents is a challenging task and event-based sampling helps resolve the issue. This book considers event-triggered scheme as a transmission protocol to negotiate information exchange in resilient control for NCS via a robust control algorithm to regulate the closed loop behavior of NCS in the presence of mismatched uncertainty with limited feedback information. It includes robust control algorithm for linear and nonlinear systems with verification.

Features:

- Describes optimal control based robust control law for event-triggered systems.

- States results in terms of Theorems and Lemmas supported with detailed proofs.

- Presents the combination of network interconnected systems and robust control strategy.

- Includes algorithmic steps for precise understanding of the control technique.

- Covers detailed problem statement and proposed solutions along with numerical examples.

This book aims at Senior undergraduate, Graduate students, and Researchers in Control Engineering, Robotics and Signal Processing.

Event-Triggered Transmission Protocol in Robust Control Systems

Niladri Sekhar Tripathy
Indra Narayan Kar
Kolin Paul

CRC Press
Taylor & Francis Group
Boca Raton London New York

CRC Press is an imprint of the
Taylor & Francis Group, an **informa** business

First edition published 2023
by CRC Press
6000 Broken Sound Parkway NW, Suite 300, Boca Raton, FL 33487-2742

and by CRC Press
4 Park Square, Milton Park, Abingdon, Oxon, OX14 4RN

CRC Press is an imprint of Taylor & Francis Group, LLC

© 2023 Niladri Sekhar Tripathy, Indra Narayan Kar and Kolin Paul

ISBN: 978-1-032-13525-0 (hbk)
ISBN: 978-1-032-13526-7 (pbk)
ISBN: 978-1-003-22969-8 (ebk)

DOI: 10.1201/9781003229698

Typeset in Nimbus font
by KnowledgeWorks Global Ltd.

Contents

SECTION I   Control of Linear Systems

SECTION III Applications

List of Figures

List of Tables

Notations

SYMBOLS AND MAIN VARIABLES

	Symbol and main variable
x	state of continuous-time system at current time
x_k	state of discrete-time system at k^{th} instant
e	measurement error
A	system matrix
B	input matrix
t_k	event-triggering instant for continuous-time system
k_i	event-triggering instant for discrete-time system
J	cost-functional
V	Lyapunov function
u	stabilizing control input
v	virtual control input
K	stabilizing controller gain
L	virtual controller gain
$\Delta A, \Delta B$	uncertain matrix
$A(p_0)$	nominal system matrix
τ	inter-event time

SPACES

	Space
$\mathbb{R}$	real line
$\mathbb{R}^n$	the n dimensional Euclidean real space
$\mathbb{R}^{n \times m}$	is a set of all $(n \times m)$ real matrices.
$\mathbb{R}_0^+$	all possible set of positive real numbers
$\mathbb{I}$	all possible set of non-negative integers

OPERATORS

	Operator
$\|x\|$	Euclidean norm of a vector $x \in \mathbb{R}^n$
$\|X\|$	norm of matrix X
$X \leq 0$	negative semi-definiteness of matrix X
$X > 0$	positive definiteness of matrix X
X^T	transpose of matrix X
X^{-1}	inverse of matrix X
$\wedge$	logical and operation
inf	infimum

ABBREVIATIONS

Abbreviation	
ADP	Approximate Dynamic Programming
CPS	Cyber-Physical System
CT-ETC	Continuous-time Event-triggered Control
CT-HJB	Continuous-time Hamilton-Jacobi-Bellman
DLE	Differential Lyapunov Equation
DT-ETC	Discrete-time Event-triggered Control
DT-HJB	Discrete-time Hamilton-Jacobi-Bellman
ETC	Event-triggered Control
ET-HJB	Event-triggered Hamilton-Jacobi-Bellman
FT-ETC	Finite-time Event-triggered Control
GHJB	Generalized Hamilton-Jacobi-Bellman
HJB	Hamilton-Jacobi-Bellman
ISS	Input-to-State Stability
LQR	Linear Quadratic Regulator
NCS	Networked Control Systems
NN	Neural Network
PDE	Partial Differential Equation
SDC	State Dependent Coefficient
SDRE	State Dependent Riccati Equation
ZOH	Zero-Order Hold

Preface

Controlling uncertain networked control systems (NCS) with limited communication among subcomponents is a challenging task. Usually in NCS, multiple physical systems interact with their subcomponents through shared communication resources. Therefore, effective utilization of these resources is the primary requirement for accomplishing the desired goal of controlled system. This fact motivates several researchers toward aperiodic sensing and control beyond the conventional continuous and periodic scheme. In the recent past, it is shown that aperiodic sampling has more benefits over periodic sampling in reducing the consumption of resources. Nowadays, the event-based sampling impressively exhibits the effective reduction of network bandwidth within the feedback loop. This book considers event-triggered scheme as a transmission protocol to negotiate information exchange in resilient control for NCS. In event-triggered control, a new information is exchanged among the cyber components only when truly needed.

Mainly in NCS, the network-related constraints like delay in the communication medium, packet drop, and single-packet versus multiple-packet transmission primarily affect control performance. Apart from communication constraint in feedback loop, the presence of system uncertainties deteriorates the closed loop performance. Mainly parameter variation, disturbances, and unmodeled dynamics are the sources of uncertainties. Broadly, system uncertainties are divided into two classes namely matched and mismatched ones. In matched system, the uncertainty belongs to the range space of input matrix. This condition does not hold for a mismatched system. The book presents a robust control algorithm to regulate the closed loop behavior of NCS in the presence of matched and mismatched uncertainties with limited feedback information. An optimal control approach for robust controller design framework is used to derive the control law. The essential idea of proposed robust control approach is an optimal control input computed for the nominal system, which minimizes a certain cost-functional. The derived optimal input for the nominal system is the robust solution of the original uncertain system. The control law is computed and actuated only when a predefined event condition is satisfied. The ISS-based analysis is used to derive the event-triggering condition and stability results.

The main results of this monograph are presented in three parts. In Part I, a robust control law is designed for a linear uncertain system. To stabilize such uncertain system, the control input is actuated aperiodically based on a predefined event-condition. The analysis is done in both continuous and periodic time-domain. Based on the results reported in Part I, Part II considers a nonlinear system for analysis. The various applications of proposed control law on different class of systems are stated in Part III. A brief description on individual chapters are outlined next. Chapter 2 describes both static and dynamic event-triggered control law for a continuous-time linear uncertain system. In this formulation, a mismatched parametric uncertainty is considered for analysis. The robust-optimal control method is adopted to derive the control law which can withstand the uncertainty introduced into the system. The similar problem for discrete-time system is solved in Chapter 3.

Chapter 4 proposes a framework to control a class of nonlinear system with limited feedback. A finite-time suboptimal event-triggered control law is derived in this chapter. The control law is computed without solving the HJB equation. A periodic robust control law for discrete-time mismatched nonlinear is proposed in Chapter 5. The results reported in this chapter ensure the asymptotic convergence of system states. The proposed linear event-triggered control methodology developed in Chapter 2 has been used in Chapter 6 to stabilize different class of nonlinear system with an aperiodic feedback information. The robust control approach considers the system nonlinearity as a source of uncertainty. The additional results are reported in appendix in order to compute the intermediate steps. The main purpose of writing this book is for the researchers and postgraduate students (PHD, Master degree) with interest in NCS. The presentation of this book is mostly self-contained.

Author Biography

Niladri Sekhar Tripathy received the bachelor's degree in electronics and communication engineering and the master's degree in mechatronics engineering in 2009 and 2011, respectively, and the Ph.D. degree in electrical engineering from IIT Delhi, in 2017. After completion of master's degree, he has worked as a Senior Research Fellow (SRF) with IIT Delhi, from 2011 to 2012. From 2018 to 2019, he has worked as a Postdoctoral Researcher with the Singapore University of Technology and Design (SUTD). He is currently working as an Assistant Professor with IIT Jodhpur, India. His research interests include control techniques for cyber-physical systems, cybersecurity, mechatronics, and application of control theory in computing.

Indra Narayan Kar received the B.E. degree in electrical engineering from the Bengal Engineering College (currently IIEST), Shibpur, India, in 1988, and the M.Tech. and Ph.D. degrees in electrical engineering from IIT Kanpur, Kanpur, India, in 1991 and 1997, respectively. From 1996 to 1998, he was a Research Student with Nihon University, Tokyo, Japan, under the Japanese Government Monbusho Scholarship Program. He joined the Department of Electrical Engineering, IIT Delhi, New Delhi, India, in 1998, where he is currently a Professor and the Institute Chair Professor. He has published over 150 articles in international journals and conferences. His current research interests include nonlinear control, time- delayed control, incremental stability analysis, cyber-physical systems, and application of control theory in power networks and robotics.

Kolin Paul is a Professor in the Department of Computer Science and Engineering at IIT Delhi India. He received his B.E. degree in Electronics and Telecommunication Engineering from NIT Silchar in 1992 and Ph.D. in Computer Science in 2002 from BE College (DU), Shibpore. During 2002-3 he did his post-doctoral studies at Colorado State University, Fort Collins, USA. He has previously worked at IBM Software Labs. His last appointment was as a Lecturer in the Department of Computer Science at the University of Bristol, UK. He has also held a Visiting Position at KTH, Stockholm. His research interests are in understanding high performance architectures and compilation systems. In particular he works in the area of Adaptive/Reconfigurable Computing trying to understand its use and implications in embedded systems. He is also involved in the design of systems for affordable healthcare.

1 Introduction

After studying this chapter, one should be able to: define networked control systems (NCS) and their issues; discuss event-triggered control technique: modeling, controller, and event-triggering law design; derive linear-quadratic-regulator (LQR)-based optimal controller for both continuous and discrete-time systems; define optimal control approach for robust controller design using LQR framework; discuss the concept of input-to-state stability (ISS) theory for continuous and discrete-time system; state different types of event-triggering laws.

1.1 NETWORKED CONTROL SYSTEM

Advances in Integrated circuit technology has created enormous communication and computation capability for NCS. Generally, in NCS, multiple physical systems are interconnected and exchange their local information using a shared digital network. Primarily, NCS have four subcomponents namely physical system, sensor, communication network, and actuator. The interconnection of each subcomponent is shown in Fig 1.1. The lower installation cost, easier to debugging, etc. are the key benefits

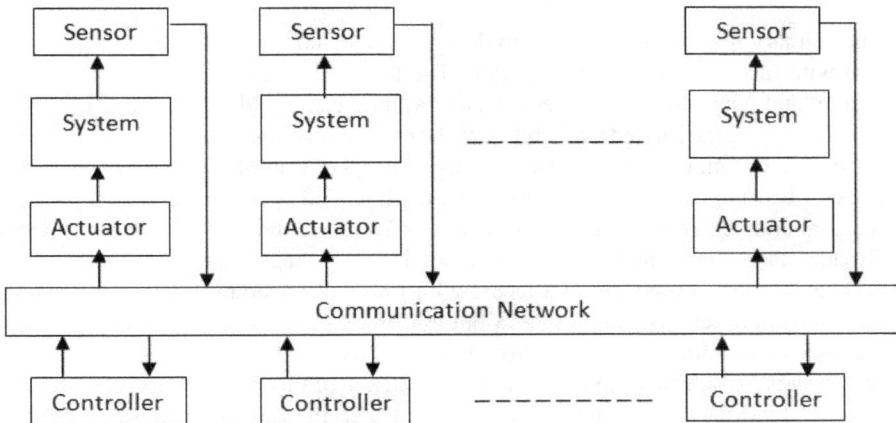

Figure 1.1 Conceptual block diagram of NCS

of NCS over the conventional control technique. But the presence of transmission medium within the feedback loop inherits several network related uncertainties like

DOI: 10.1201/9781003229698-1

delay in the communication medium, packet drop, and single-packet versus multiple-packet transmission which primarily affect the closed loop performance and stability. Due to shared nature of the communicating medium, the continuous or periodic transmission of information incurs a large bandwidth requirement. Apart from bandwidth requirement, most NCS are powered by battery sources which make it imperative to efficiently use energy. It has been observed that power consumption is directly proportional to the amount of data that is transmitted over the network. The possibility of a trade-off between communication, computation, and power consumption has motivated a large number of researchers to pursue research on NCS with limited sensing and actuation [13, 66, 146, 58]. Recently, an event-triggered-based control technique has been proposed in [117, 40], to reduce the information requirement for realizing a stable control law. Several researchers have used the event-triggered-control algorithm in different practical applications such as robotics [104, 124, 62], power system [138, 20, 36], filtering and estimation problem [137, 123, 78], communication protocol designing [122, 99], control of multi-agent system [32, 42, 97], wireless sensor network [35, 145], etc.

Apart from communication constraints, the unmodeled system dynamics also plays a major role in system instability. To tackle system uncertainty in presence of communication constraints, this book proposes a robust sporadic feedback policy between the system and controller such that the closed loop system remains stable. The control law and sporadic feedback policy are derived based on the optimal control method and input-to-state stability (ISS) theory, respectively.

The main focus of this monograph is stated in next subsection.

1.2   FOCUS OF THIS BOOK

The primary focus of this book is to design a robust controller for an uncertain system with limited feedback information. The feedback information is limited as the system and controller are connected using a finite bandwidth communication channel. To capture the limited availability of the communication channel, an event-based control algorithm is proposed. The detailed concept of event-based control algorithm is described in next section. In this book, the bounded variation of system parameters and unmodeled system dynamics are considered as the sources of uncertainties. Broadly, the system uncertainty is subdivided in two categories namely matched and mismatched one. A system is affected with matched uncertainty if the uncertainty is expressed in terms of input matrix but this assumption does not hold for the mismatched case. Although the stability of matched system has been reported by many researchers, it is difficult to ensure the convergence of mismatched uncertain system analytically. This book proposes novel event-triggering rule and controller to stabilize a mismatched linear and nonlinear systems with limited feedback information. The ISS theory is used to design the event-triggering law.

Apart from proposing an event-based robust controller, this book also presents a novel finite-time event-triggered control law for a class of nonlinear system where an approximate solution of Hamilton-Jacobi-Bellman (HJB) equation is used for deriving a nonlinear control law. To derive the control law, the HJB equation is approximated as a state-dependent differential Riccati equation (SDRE). The designed event-triggered control law is easy for on-line implementation and it also ensures the ISS of the closed loop system.

1.3 EVENT-TRIGGERED CONTROL

In event-triggered control, sensing and actuation at the system end happens only when a pre-specified event-condition is violated. As such, the event-triggered control algorithm is subdivided into two primary parts. Firstly, the feedback controller which computes the control input for the physical system depends on the transmitted state or output and secondly, a predefined event-triggering condition, which decides the next time instant for sensing and actuation. This event-condition mostly depends on the current states or outputs of the system. In event-triggered control, the control input is actuated aperiodically. A zero-order-hold (ZOH) is used to hold the last transmitted control input until the next input is transmitted. To hold the last transmitted control input at actuator end, the application of ZOH is very common in event-triggered control due to its simplicity. A simple Digital to Analog converter (D/A converter) is used to reconstruct the transmitted signal. A conceptual block diagram of event-triggered control technique is shown in Fig 1.2. The primary shortcoming

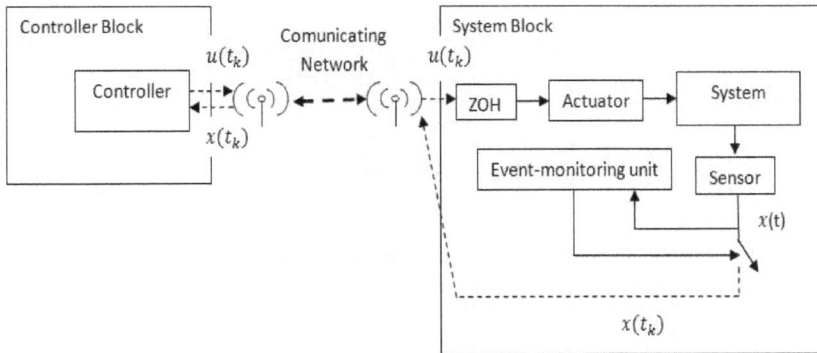

Figure 1.2 Conceptual block diagram of event-triggered control. Dotted lines represent the aperiodic information transmission through the communication channel. The symbols x and u denote the system state and control input, respectively.

of continuous-time event-triggered control is that it requires a continuous monitoring of event-condition. In [57, 56], Heemels et al. have proposed an event-triggering technique where event-condition is monitored periodically. To avoid continuous or periodic monitoring, self-triggered control technique has been reported in [9, 134] where the next event occurring instant is computed analytically using system's state at the previous sampling instant. The name self-triggered is used as the controller itself decides the next transmission instant and closes the feedback loop for sensing and actuation at the system end. In self-triggered control, the closed loop performance and stability analysis are done along the same line as in the event-triggered control case. Therefore self-triggered control technique can be treated as a special case of event-triggered control approach.

Remark 1 *The time-triggered control is an alternative actuation process of event-triggered control aiming to reduce the communication resources. In this approach, the control inputs are actuated with a periodic time-interval. The actuation instants are decided based on a clock signal. A block diagram of time-triggered control is shown in Fig. 1.3. Generally, the results of sampled-data control are used for analysis of such systems.*

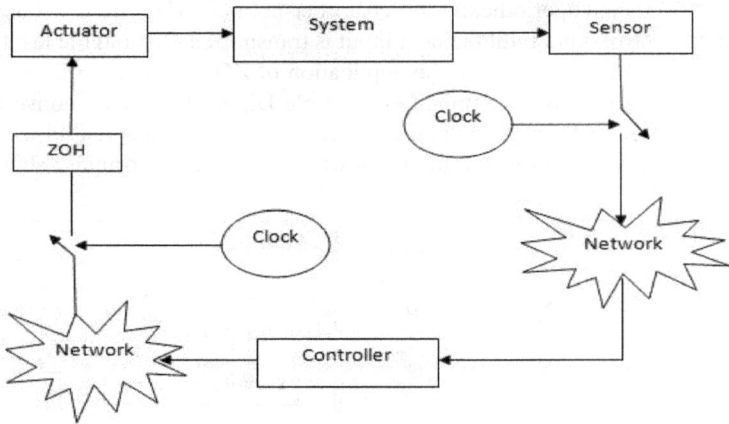

Figure 1.3 Conceptual block diagram of time-triggered control

1.3.1 MODELING

Consider a linear continuous-time system with a state feedback control input as

$$\begin{cases} \dot{x}(t) = Ax(t) + Bu(t) \\ u(t) = Kx(t) \end{cases} \tag{1.1}$$

where $x \in \mathbb{R}^n$ and $u \in \mathbb{R}^m$ are system state and input vector, respectively. The matrix K is controller gain. In event-triggered control, the control input $u(t)$ is actuated aperiodically. Due to aperiodic actuation, the system (1.1) is written as

$$\begin{cases} \dot{x}(t) = Ax(t) + Bu(t_k), t \in [t_k, t_{k+1}) \\ u(t_k) = Kx(t_k) \end{cases} \tag{1.2}$$

where t_k and $x(t_k)$ are the time-instant at which kth event is occurred and state information at t_k, respectively.

Remark 2 *The implementation of a control input on an embedded processor can be considered as a job, which consists of receiving the state information from the system, executing and actuating the control law. It is assumed that the sensing at system end, computation at controller end, and actuation at actuator end are happening instantaneously. Based on Fig. 1.4, let t_k is the time-instant at which kth control task is released. Now the elapsed time in between two consecutive events is denoted by τ and defined as $\tau = t_{k+1} - t_k$. The control input $u(t_k)$ remains fixed as it is until the next event occurring instant t_{k+1}. In the conventional control scheme, the numerical value of τ is constant and pre-specified. But in event-triggered control, τ is changing aperiodically based on the system dynamics.*

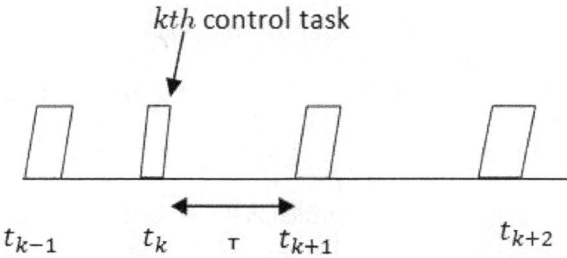

Figure 1.4 Inter-event time, τ

Several researchers have modeled the event-triggered system (1.2) as an impulsive system, perturbed system, sampled data system, and time-delay system. A perturbed system modeling strategy is adopted in this book, which is briefly discussed. In [52], P. Tabuada modeled the event-triggered system (1.2) as a continuous-time perturbed system by introducing an another variable $e(t)$. The variable $e(t)$ is referred as measurement error and it is defined as

$$e(t) = x(t_k) - x(t), t \in [t_k, t_{k+1}). \qquad (1.3)$$

Using (1.3), the event-triggered system (1.2) is written as

$$\dot{x}(t) = Ax(t) + BK\{x(t) + e(t)\}. \qquad (1.4)$$

In [40], the similar approach is adopted to model a discrete-time event-triggered system, which is defined as

$$x(k+1) = Ax(k) + BK\{x(k) + e(k)\}, k \in [k_i, k_{i+1}) \qquad (1.5)$$

where $e(k) = x(k_i) - x(k)$ and $u(k_i) = K\{x(k) + e(k)\}$. Here k_i represents the last event-occurring instant and the input is actuated at an aperiodic discrete instant k_0, k_1, $k_2 \cdots k_i$, where $i \in 1, 2, 3 \cdots \mathbb{N}$. The ISS theory has been used to analyze the closed-loop stability and performance of (1.4) and (1.5).

1.3.2 CONTROLLER DESIGN

The presence of communication network within the feedback loop complicates the controller design processes. In event-triggered control, there are several controller design processes. Brief descriptions on different controller design techniques are discussed below:

(i) Emulation-based approach: In this approach [116, 55], initially the controller is designed in continuous-time domain ignoring the network effect. Then the sufficient conditions are derived under the presence of communication network. This book adopts the emulation-based approach to design the controller under limited channel bandwidth.

(ii) Co-design-based approach: The co-design approach [31, 108, 43] considers the network properties and controller design issues simultaneously. The control law is changed depending on the variation of network issues. This approach is comparatively harder than the emulation-based approach.

(iii) Discrete method: In this approach, a discrete-time model is computed from a continuous system model, adopting the discretization method [23, 125]. Then, based on the discrete-time model of system, a stable controller is designed. The control law is verified in the presence of communication network and sufficient conditions are derived [94, 27, 45]. The reconstruction of an exact discrete-time model of a continuous-time nonlinear system is very difficult. Therefore, this method is mostly applicable for linear systems.

1.3.3 TRIGGERING CONDITION

The design of event-triggering condition is the key part of event-triggered control. Basically it ensures the closed loop stability and the existence of positive inter-event time, which helps to avoid the Zeno behavior[1]. Figure 1.5 shows that the event is occurred at t_k instant only when $\|e(t)\|$ touches the threshold limit. It also shows that the inter-event time τ is not a constant parameter, it varies with respect to time t. Different techniques have been used in past studies to design an event-triggering condition. Some of them are discussed below:

Constant threshold: The design of event-triggering condition depends on the growth of $\|e(t)\|$. The definition of $e(t)$ is defined in (1.3). Through an event-triggering rule, the control input is computed only when $\|e(t)\|$ exceeds a predefined constant threshold. That means,

$$\|e(t)\| \geq \Xi \qquad (1.6)$$

---

[1] Infinite number of transmission and computation in a finite time [69].

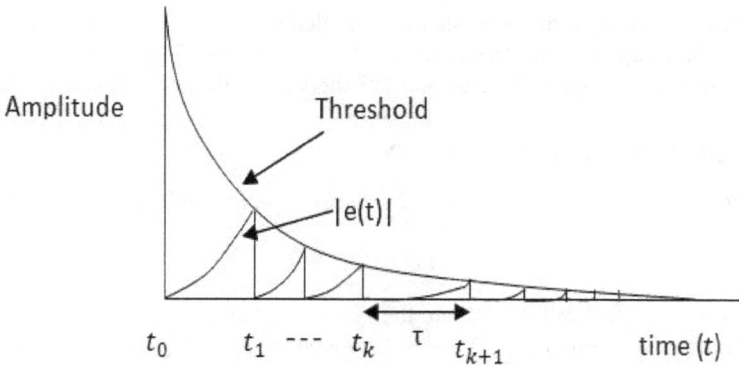

Figure 1.5 The design of event-triggering condition

where the constant $\Xi > 0$ is a design parameter. This control technique is also known as deadband control [74, 98].

State-dependent threshold: In [117, 40], the event condition is defined based on the systems state. In this book, we have adopted this approach to define an event-triggering condition. The event condition is derived from the ISS theory. In general, the state dependent event-condition for (1.4) is expressed as

$$\|e(t)\| \geq \sigma \|x(t)\| \tag{1.7}$$

where the design parameter $\sigma \in (0,1)$. The triggering condition (1.7) ensures the stability of closed loop system (1.4).

Dynamic threshold: Maximizing inter-event time is the key aim of the event-triggered or self-triggered control in order to reduce the total communication requirement. A. Girad [52] has proposed a new event-triggering mechanism called dynamic event-triggering to achieve larger inter-event time with respect to the previous approach [117]. In dynamic event-triggering mechanism, a variable η is used along with the system state to realize an event-condition. To define the dynamic variable $\eta(t)$, an extra differential equation is needed. Now to analyze the stability of x, the dynamics of η plays an impotent role. Therefore defining the dynamics of η ($\dot{\eta}(t)$) is a challenging problem and solving $\dot{\eta}(t)$ numerically also increases the computation burden.

1.4   BACKGROUND

This section introduces the preliminary results of optimal control theory for both linear and nonlinear system. Based on optimal control theory, a robust controller is designed where system model is affected by bounded parametric uncertainty. Apart from controller design, the analysis of controlled uncertain system is carried out by adopting the Lyapunov and ISS theory. A brief overview of optimal control theory, robust controller design technique, and ISS theory are discussed in next subsections.

1.4.1 OPTIMAL CONTROL THEORY

Optimal control for continuous-time system: Consider a nonlinear system as

$$\dot{x}(t) = f(x) + g(x)u(t) \tag{1.8}$$

where vectors $x \in \mathbb{R}^n$ and $u \in \mathbb{R}^m$ are the system sate and input vectors, respectively. The nonlinear functions $f(x) : \mathbb{R}^n \to \mathbb{R}^n$ and $g(x) : \mathbb{R}^n \to \mathbb{R}^{n \times m}$ are smooth mappings and $x = 0$ is the equilibrium point e.g., $f(0) = 0$ and $g(0) = 0$. To stabilize (1.8), an optimal control input u^* is designed by minimizing the cost-functional

$$J(x,u) = \int_0^\infty (x^T Q x + u^T R u) dt. \tag{1.9}$$

Control objective: Design a state feedback control law $u^*(x)$ for (1.8), in order to minimize (1.9).

To obtain the above mentioned objective, following lemma is introduced [91, 72, 7].

Lemma 1 *Suppose $J^*(x,u)$ represents the minimal cost of (1.9) and an optimal control input $u^* = -R^{-1} B^T \left(\frac{\partial J^*}{\partial x} \right)$ for (1.8) is designed, which minimizes the cost-functional (1.9). Then, the optimal input u^* satisfies the following HJB equation*

$$-\frac{\partial J^*}{\partial t} = min_{u(t) \in \mathbb{R}^m} \left\{ x^T Q x + u^{*T} R u^* + \left(\frac{\partial J^*}{\partial x} \right)^T f(x,u^*) \right\}. \tag{1.10}$$

Now, for a linear system (1.1) and cost-functional (1.9), the HJB equation (1.10) reduces to following Riccati equation

$$PA + A^T P - PBR^{-1} B^T P + Q = 0. \tag{1.11}$$

The positive definite solution P of (1.11) is used to compute the optimal input for (1.1) as

$$u^* = -R^{-1} B^T P x(t). \tag{1.12}$$

Optimal control for discrete-time system: Let an affine-in-input nonlinear discrete-time system is described as

$$x_{k+1} = f(x_k) + g(x_k)u_k \tag{1.13}$$

where x_k and u_k are system state and input vectors, respectively. The functions $f(x_k)$ and $g(x_k)$ are smooth and $x = 0$ is an equilibrium point e.g., $f(0) = 0$ and $g(0) = 0$.

To stabilize (1.13), it is essential to design a control input $u_k \in \mathbb{R}^m$, which minimizes the following quadratic cost-functional

$$J(x_k, u_k) = \sum_{k=0}^{k=\infty} (x_k^T Q x_k + u_k^T R u_k) \tag{1.14}$$

where matrices $Q \geq 0$ and $R > 0$.

Control objective: Design a state feedback control law $u^*(x_k)$ for (1.13), in order to minimize (1.14).

To satisfy the above mentioned objective, a Lemma is introduced below:

Lemma 2 *Suppose $J^*(x_k, u_k)$ represents the minimal cost of (1.14) and an optimal control input $u_k^* = -\frac{1}{2}R^{-1}g^T(x_k)\frac{\partial J^*(x_{k+1})}{\partial x_{k+1}}$ for (1.13) is designed, which minimizes the cost-functional (1.14). Then, the optimal input u_k^* satisfies the following HJB equation:*

$$J^*(x_k, u_k) = x_k^T Q x_k + u_k^{*T} R u_k^* + \sum_{n=k+1}^{\infty} \left(x_n^T Q x_n + u_n^{*T} R u_n^* \right). \tag{1.15}$$

For a linear system

$$x_{k+1} = A x_k + B u_k \tag{1.16}$$

and a cost-functional (1.14), the HJB equation (1.15) transforms to a discrete-time Riccati equation as

$$A^T \{P^{-1} + BR^{-1}B^T\}^{-1} A - P + Q = 0. \tag{1.17}$$

The positive-definite solution P of (1.17) is used to compute the optimal input as

$$u_k^* = -\frac{1}{2}R^{-1}B^T(P^{-1} + BR^{-1}B^T)^{-1} A x_k. \tag{1.18}$$

The proof of these Lemmas 1 and 2 are included in Appendix B (Page No. 151 – 153).

Remark 3 *Solving HJB equation (1.10) [or (1.15)] is computationally expensive as it is a partial differential equation. This is the primary drawback in nonlinear optimal control problem. In this book, neural network (NN)-based least square approach is used to approximate the solution of DT-HJB equation which is discussed in Chapter 5. On the other hand, many researchers propose the optimal control theory for a class of nonlinear system where function $f(x_k)$ has been rewritten as $f(x_k) = A(x_k)x_k$. That means, the system equation (1.13) is expressed by a linear like equation where system matrix A has state dependency. This representation is called as state dependent coefficient form [28] and it helps to convert HJB into a state-dependent Riccati equation (SDRE), which is an ordinary matrix differential equation.*

1.4.2 ROBUST CONTROL DESIGN

In the previous subsection, the optimal control results for both continuous and discrete-time system are briefly mentioned. F. Lin et al. [81, 79], have used the optimal control theory to design the robust controller for an uncertain system. They have considered that the uncertainty is affecting the system model due to the bounded variation of system parameters.

Robust control problem: Suppose an uncertain system is described as

$$\dot{x}(t) = A(p)x(t) + Bu(t) \tag{1.19}$$

where $p \in \hat{P}$ is an unknown parameter vector and upper bounded by the known set $\hat{P}$. Initially, the matching condition is considered for study. For a matched system, the uncertainty is in the range space of input matrix B. That means, the system matrix $A(p)$ can be expressed as $A(p) - A(p_0) = B\Phi(p)$ for some unknown matrix $\Phi(p)$, where $p_0 \in \hat{P}$ is the nominal value of p. Here, the matrix $A(p_0)$ is the nominal part of uncertain system matrix $A(p)$. The uncertain matrix $\Phi(p)^T\Phi(p)$ is upper bounded by a known matrix F and it satisfies the following inequality

$$\Phi(p)^T\Phi(p) \leq F, \ \forall p \in \hat{P}. \tag{1.20}$$

The primary goal of the present robust control problem is, design a state feedback control law $u = Kx(t)$ which makes closed loop system (1.19) stable. To design the robust controller gain K, an equivalent optimal control problem is formulated. The optimal control problem is solved based on the nominal dynamics by minimizing a quadratic cost-functional. A brief description on optimal control approach for robust controller design method is discussed next.

Optimal control based solution: Suppose the uncertain system (1.19) has a nominal dynamics as

$$\dot{x}(t) = A(p_0)x(t) + Bu(t). \tag{1.21}$$

To formulate an equivalent optimal control problem, following cost-functional is selected

$$J(x, u) \quad = \quad \int_0^\infty (x^T F x + x^T Q x + u^T R u) dt. \tag{1.22}$$

The essential idea of optimal control approach for robust controller design is, $\forall p \in \hat{P}$, the optimal input for (1.21) which minimizes the modified[2] cost-functional (1.22) is the robust solution of the original uncertain system (1.19).

In Chapter 2, the concept of robust controller design using optimal control framework is discussed in detail. Throughout this book, the optimal control based solution is used to design the robust controller for both continuous and discrete-time uncertain systems.

---

[2]Usually, we adopt quadratic cost-functional of the form (1.9). However, here we will add an additional term $x^T F x$ in order to reflect the uncertainties. Hence, it will be referred as modified cost-functional.

1.4.3   STABILITY RESULTS

To analyze the stability of closed loop system, the book use the Lyapunov and ISS theories. Before introducing the stability results, following definitions are stated [71].

Definition 1 *A scalar function $V(x)$ is considered as a*

  i) *positive-definite function, if $V(x) > 0$ within the region $\|x\| \leq \varepsilon$ except at the origin, where $V(0) = 0$;*
  ii) *negative-definite function, if $[-V(x)]$ is positive definite;*
  iii) *positive semi-definite function, if $V(x) \geq 0$ within the region $\|x\| \leq \varepsilon$ including the origin, where $V(0) = 0$;*
  iv) *negative semi-definite function, if $[-V(x)]$ is positive semi-definite.*

Definition 2 (Class $\mathscr{K}$ function) *Function $\alpha(r) : \mathbb{R}_{\geq 0} \to \mathbb{R}_{\geq 0}$ is called as a class $\mathscr{K}$ function if it is continuous and strictly increasing and holds $\gamma(0) = 0$.*

Definition 3 (Class $\mathscr{K}\infty$ function) *The function β is a Class $\mathscr{K}\infty$ function, if it is a Class $\mathscr{K}$ function and $\beta(r) \to \infty$ for $r \to \infty$.*

Definition 4 (Class $\mathscr{K}\mathscr{L}$ function) *The function $\gamma(r,s) : \mathbb{R}_{\geq 0} \times \mathbb{R}_{\geq 0} \to \mathbb{R}_{\geq 0}$ is a Class $\mathscr{K}\mathscr{L}$ function, if it is a Class $\mathscr{K}$ function with respect to r for fixed s and it is strictly decreasing with respect to s when r is fixed.*

Below, we first introduce the Lyapunov theory.

Definition 5 *Consider a state equation*

$$\dot{x} = f(x), f(0) = 0 \tag{1.23}$$

where x is the state vector. Let there exist a scalar $\varepsilon > 0$ and a continuous and smooth positive-definite function $V(x)$, $\forall \|x\| \leq \varepsilon$. The equilibrium point $x^e = 0$ of (B.6) is

  i) *asymptotically stable if $\dot{V}(x)$ is negative definite $\forall x \neq 0$ along the trajectory of (B.6) or if $\dot{V}(x)$ is negative semi-definite and no trajectory can remain on the line at which $\dot{V}(x) = 0$ forever, except the origin;*
  ii) *stable in-the-sense of Lyapunov if $\dot{V}(x) = 0$ along the trajectory of (B.6).*

Discrete-time system: The Lyapunov stability theorem is also extended for a discrete-time system which is discussed next:

Definition 6 *Suppose there exists a Lyapunov function $V(x(k))$ for a discrete-time system*

$$\dot{x} = f(x(k)), \; f(0) = 0. \tag{1.24}$$

The origin of (1.24) is asymptotically stable along the state trajectory x if $\Delta V(x(k)) = [V(x(k+1)) - V(x(k))] < 0, \; \forall x \neq 0.$

Next, the concept of input-to-state stability is described briefly.

1.4.3.1   Input-to-state Stability for Continuous-time System

Consider a state space representation of a continuous time system as

$$\dot{x} = f(t, x, u) \tag{1.25}$$

where function f is continuous in t and locally Lipschitz in u and x respectively. The unforced system of (1.25) is

$$\dot{x} = f(t, x, 0) \tag{1.26}$$

which has an equilibrium point at $x = 0$. For a linear system

$$\dot{x} = Ax + Bu \tag{1.27}$$

the solution $x(t)$ is

$$x(t) = e^{(t-t_0)A} x(t_0) + \int_{t_0}^{t} e^{(t-\tau)A} Bu(\tau) d\tau \tag{1.28}$$

where A is Hurwitz matrix and t_0 is the initial time instant. Using (1.28), the upper-bound of x is computed as

$$
\begin{aligned}
\|x(t)\| \;\leq\;& ke^{-\lambda(t-t_0)} \|x(t_0)\| + \int_{t_0}^{t} ke^{-\lambda(t-t_0)} \|B\| \|u(\tau)\| d\tau \\
\leq\;& ke^{-\lambda(t-t_0)} \|x(t_0)\| + \frac{k\|B\|}{\lambda} \sup_{(t_0 \leq \tau \leq t)} \|u(\tau)\|
\end{aligned}
\tag{1.29}
$$

where $\|e^{(t-t_0)A}\| \leq ke^{-\lambda(t-t_0)}$. The equation (1.29) ensures the exponential convergence of state x for $u = 0$. Moreover, (1.29) shows that a bounded-input-bounded-state property. However this property does not hold for a nonlinear system even if the the origin of (1.26) is globally uniformly asymptotically stable.

Let us consider (1.25) as a perturbed system of (1.26) and holds the following property

$$\|f(t, x, u) - f(t, x, 0)\| \leq L\|u\|, \; \forall t > t_0. \tag{1.30}$$

Suppose $V(t,x)$ is a Lyapunov function for (1.26). Now, in the presence of control input u, it is possible to show that the $\dot{V}(x)$ is negative outside a ball of radius r. Here r is bounded by $sup\|u\|$. This can be shown that the state x is bounded by

$$\|x(t)\| \le \beta(\|x(t_0)\|,t) + \gamma(sup\|u(\tau)\|) \tag{1.31}$$

where β and γ are KL and K functions, respectively. The Fig. 1.6 shows that as time $t \to \infty$ the state trajectories $x(t)$ reaches to the circle having a radius γ.

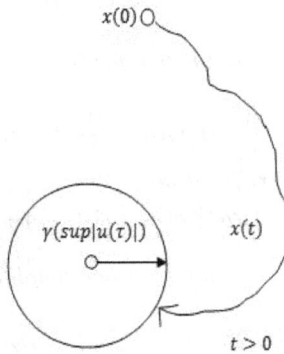

Figure 1.6 The trajectory of x under ISS property

This motivates to state the following definition of input-to-state stability.

Definition 7 *Suppose there exist a KL function β and a K function γ such that the state of the system (1.25) is bounded by*

$$\|x(t)\| \le \beta(\|x(t_0)\|,t) + \gamma(sup\|u(\tau)\|) \tag{1.32}$$

for any selection of initial state $x(t_0)$ and bounded input $u(t)$. Then the system (1.25) is called as input-to-state stable with respect to control input u.

Observation: The following observations are made from (1.32):

for any bounded $u(t)$, the state $x(t)$ will be bounded.
if time t increases, the $x(t)$ will be bounded by $\gamma(sup\|u(\tau)\|)$.
for (1.26), the $x(t) \to 0$ as $t \to \infty$. This is possible as

$$\|x(t)\| \le \beta(\|x(t_0)\|,t) \tag{1.33}$$

In the following, a Lyapunov like theorem is introduced to derive the sufficient condition for ISS.

Definition 8 *A continuous function $V(x) : \mathbb{R}^n \to \mathbb{R}$ is an ISS Lyapunov function for (1.25) if there exist class k_∞ functions $\alpha_1, \alpha_2, \alpha_3$, and γ for all $x, d \in \mathbb{R}^n$ and it satisfy*

$$\alpha_1(\|x(t)\|) \leq V(x(t)) \leq \alpha_2(\|x(t)\|) \tag{1.34}$$

$$\nabla V(x)\dot{x} \leq -\alpha_3(\|x(t)\|) + \gamma(\|u(t)\|) \tag{1.35}$$

Then the system (1.25) is ISS with respect to input $\gamma(\|u(t)\|)$.

Proof *Let $V(x)$ is a Lyapunov function for (1.25) and $\dot{V}(x)$ with respect to x is*

$$
\begin{aligned}
\dot{V} &= \frac{\partial V}{\partial t} + \frac{\partial V}{\partial x} f(t,x,0) + \frac{\partial V}{\partial x}[f(t,x,u) - f(t,x,0)] \\
&\leq -\alpha_3(\|x\|) + \gamma(\|u\|) \\
&\leq -(1-\theta)\alpha_3(\|x\|) - \theta\alpha_3(\|x\|) + \gamma(\|u\|)
\end{aligned}
\tag{1.36}
$$

where parameter $\theta \in (0,1)$. Now, after further simplification, the above inequality reduces to

$$\dot{V} \leq -(1-\theta)\alpha_3(\|x\|), \ \forall \ \|x\| \geq \alpha_3^{-1}\left(\frac{\gamma(\|u\|)}{\theta}\right) \tag{1.37}$$

for all (t,x,u). Here $(1-\theta)\alpha_3(\|x\|)$ is a class K function.

An example is provided to illustrate the above results.
Example: Consider a nonlinear system as

$$\dot{x} = -x^3 + u \tag{1.38}$$

The unforced system of (1.38) has a globally asymptotically stable origin. Let $V(x) = \frac{x^2}{2}$ is an ISS Lyapunov function for (1.38), then the $\dot{V}(x)$ along the x is

$$
\begin{aligned}
\dot{V} &= -x^4 + xu \\
&= (1-\theta)x^4 - \theta x^4 + xu \\
&\leq -(1-\theta)x^4, \ \forall \ \|x\| \geq \left(\frac{\|u\|}{\theta}\right)^{1/3}
\end{aligned}
\tag{1.39}
$$

where $\theta \in (0,1)$. The system (1.38) is ISS for $\gamma(r) = \left(\frac{r}{\theta}\right)^{1/3}$.

1.4.3.2   ISS for Discrete-time System

ISS theory is also applicable for discrete-time system and following definitions are used to state the concepts:

Definition 9 *A system*

$$x(k+1) = Ax(k) + Bu(k) \tag{1.40}$$

is globally ISS if it satisfies

$$\|x(k)\| \leq \beta(\|x(0)\|, k) + \gamma\left(sup_{\tau \in [0,\infty)}\{\|u(\tau)\|\}\right) \tag{1.41}$$

with each input $u(k)$ and each initial state $x(0)$. The functions β and γ are $\mathscr{KL}$ and $\mathscr{K}_\infty$ functions, respectively.

Definition 10 *Suppose origin is an equilibrium point of a discrete-time system $x(k+1) = f(x(k), u(k))$, i.e., $f(0,0) = 0$, $\forall\, k > 0$. A positive continuous function $V(x(k)) : \mathbb{R}^n \to \mathbb{R}$ is an ISS Lyapunov function for this system if there exist class $\mathscr{K}_\infty$ functions $\alpha_1, \alpha_2, \alpha_3$ and a class $\mathscr{K}$ function γ for all $x \in \mathbb{R}^n$ and $u \in \mathbb{R}^m$ which satisfy the following conditions.*

$$\alpha_1(\|x(k)\|) \leq V(x(k)) \leq \alpha_2(\|x(k)\|), \tag{1.42}$$
$$V(x(k+1)) - V(x(k)) \leq -\alpha_3(\|x(k)\|) + \gamma(\|u(k)\|). \tag{1.43}$$

An example is considered in next section to illustrate the event-triggered control technique in the context of linear time-invariant (LTI) system. The control input is computed and transmitted at the actuator end only when a state dependent event-condition is satisfied. The triggering rule and the event-based controller are derived based on the concept mentioned in Section 1.4.

1.5 MOTIVATING EXAMPLE

Example 1: continuous-time event-triggered control

Consider a continuous-time LTI system (1.4) where matrices $A = \begin{bmatrix} 0 & 1 \\ 1 & 0 \end{bmatrix}$ and $B = \begin{bmatrix} 0 \\ 1 \end{bmatrix}$. The realization of event-triggered control algorithm is subdivided in two parts.

Part 1. Controller design: To design the controller gain K, emulation-based approach is adopted. That means, initially, the control law is designed for (1.1), which minimizes (1.9) excluding the event-triggering effect. An optimal control method is used to design the controller gain K. The state feedback controller gain $K(= R^{-1}B^T P)$ depends on a positive-definite matrix P, which satisfies the following algebraic Riccati equation (ARE)

$$PA + A^T P - PBR^{-1}B^T P + Q = 0. \tag{1.44}$$

Now, for a selection of design matrices $Q = \begin{bmatrix} 4 & 0 \\ 0 & 4 \end{bmatrix}$ and $R = 1$, the solution

of (1.44) is computed as $P = \begin{bmatrix} 7.23 & 3.23 \\ 3.23 & 3.23 \end{bmatrix}$. Using the matrix P, the optimal

control input $u(t) = Kx(t)$ is computed where $K = \begin{bmatrix} -3.2361 & -3.2361 \end{bmatrix}$. The controller gain K of (1.21) and aperiodic state information $x(t_k)$ of (1.4) are used to design $u(t_k)$. The design procedure of triggering condition is stated below.

Part 2. Triggering condition design: Consider an ISS Lypunov function $V(x) = x^T Px$. The time derivative of $V(x)$ along the solution of (1.4) is

$$
\begin{aligned}
\dot{V}(x) &= -x^T Q_1 x + 2x^T PBKe \\
&\leq -\frac{\lambda_{min}(Q_1)}{2} \|x\|^2 + \frac{2}{\lambda_{min}(Q_1)} \|PBK\|^2 \|e\|^2 \quad (1.45)
\end{aligned}
$$

where $Q_1 = (A_c^T P + PA_c)$, $A_c = (A + BK)$ and $\lambda_{min}(Q_1)$ is the minimum eigenvalue of matrix Q_1. The equation (1.45) ensures the ISS of (1.4). Now for a selection of following relation

$$
\|e\|^2 \leq \frac{\sigma \lambda_{min}^2(Q_1)}{4\|PBK\|^2} \|x\|^2 = \mu\|x\|^2, \ \sigma \in (0,1), \quad (1.46)
$$

the $\dot{V}(x)$ reduces to

$$
\dot{V}(x) \leq (\sigma - 1)\|x\|^2. \quad (1.47)
$$

Here σ and μ are design parameters and the numerical value of $\mu \left(= \frac{\sigma \lambda_{min}^2(Q_1)^2}{4\|PBK\|^2} \right)$ and σ are 0.07 and 0.1, respectively. Now, $\forall \sigma \in (0,1)$ the $\dot{V}(x) \leq 0$ which ensures the closed loop stability of (1.4).

Example 2: discrete-time event-triggered control

Consider the discrete-time LTI system (1.16), where matrices $A = \begin{bmatrix} 0.1 & 1.2 \\ 0.007 & 1.05 \end{bmatrix}$ and $B = \begin{bmatrix} 300 & 200 \\ 0.5 & 0.001 \end{bmatrix}$. Similar to Example 1, the event-triggered control technique is realized in two steps.

Step 1. Controller Design To design the controller, the discrete-time LQR optimal is formulated. The optimal control input minimizes the cost-function (1.14). To derive the controller gain matrices K, the solution Riccati equation (1.17) P, is computed for a selection of matrices $Q = \begin{bmatrix} 0.001 & 0 \\ 0 & 0.001 \end{bmatrix}$ and $R = \begin{bmatrix} 0.01 & 0.01 \\ 0.01 & 0.01 \end{bmatrix}$. Using $K = -\frac{1}{2}R^{-1}B^T(P^{-1} + BR^{-1}B^T)^{-1}A$, the controller gain matrix is computed as $K = \begin{bmatrix} 0.0028 & 0.3023 \\ -0.0037 & -0.4466 \end{bmatrix}$. Now multiplying controller gain K with the aperiodic state information $x(k_i)$, the event-triggered control input $u(k_i) = Kx(k_i)$ is computed. Here, k_i represents the ith event occurring instant. The event-triggering sequence $k_i, \forall i \in (1, n)$ is derived in next step.

Step 2. Triggering law Similar as Example 1, the event-triggering condition is derived from stability results. Let a Lyapunov function for (1.16) is $V(x) = x(k)^T P x(k)$, then $\Delta V = V(k+1) - V(k)$ is simplified as

$$\Delta V = [(A+BK)x+BKe]^T P[(A+BK)x+BKe] - x^T Px$$
$$= x^T[(A+bk)^T P(A+Bk) - P]x + 2x^T(A+Bk)^T PBKe + e^T K^T B^T PBKe \tag{1.48}$$

Now consider there exists a positive definite matrix Q for which following holds

$$(A+Bk)^T P(A+Bk) - P = -Q. \tag{1.49}$$

Then applying (1.49), (1.48) is simplified as

$$\Delta V \leq -x^T Qx + 2x^T(A+Bk)^T PBKe + e^T K^T B^T PBKe \tag{1.50}$$
$$\leq -x^T Qx + \varepsilon \tag{1.51}$$

After further simplification of (1.50), the following is obtained

$$\Delta V \leq \underbrace{-\frac{1}{2}\lambda_{min}(Q)\|x\|^2}_{\alpha_3(\|x\|)} + \underbrace{\left(\frac{2\|(A+Bk)^T PBK\|^2}{\lambda_{min}(Q)} + \|K^T B^T PBK\|^2\right)\|e\|^2}_{\gamma(\|e\|)} \tag{1.52}$$

The inequality (1.52) holds the property (1.43). Using (1.52), the event-triggering event-triggering condition is obtained

$$\|e\|^2 \geq \mu\|x\|^2 \tag{1.53}$$

where $\mu = \frac{\sigma\lambda_{min}^2(Q)}{4\|(A+Bk)^T PBK\|^2 + \sigma\lambda_{min}(Q)\|K^T B^T PBK\|^2}$, $\forall \sigma \in (0,1)$. To realize the event-triggering condition, the design parameter μ is computed as 0.2934 for a selection of $\sigma = 0.98$.

1.6   OVERVIEW OF THE LITERATURE

The selection of sampling period to control a feedback system plays an important role [6]. In traditional digital control [23, 125], sampling periods are generally considered with fixed interval. In 1960s, the concept of adaptive sampling period was considered by [38, 70]. Then, for an aperiodic increment of sampling time, the proposed concept described in [38] was theoretically verified in [54]. The fundamental research on the selection of aperiodic sampling time to control over finite bandwidth networked system has been reported in [41, 39, 114, 120]. It is shown in [66, 146, 121, 8] that aperiodic sampling has more benefits over periodic sampling, which motivates control researchers toward event-based control. Recently, immense research has been done on event-triggered control to reduce the information requirements for realizing a stable control law for NCS. The initial work on event-triggered control started with the results in [13] and experienced increasing concernment through [117]. On the other hand, to tackle model uncertainties in system dynamics, proposing a robust control law based on uncertainty bound is a mature research area. In this section, the existing research work related to event-triggered control and optimal control approach for robust controller design are reviewed.

1.6.1   EVENT-TRIGGERED CONTROL

Event-triggered control for linear and nonlinear system: The effective reduction of data processing while ensuring satisfactory closed loop performance through event-based control algorithm has reported in [117, 84, 140, 118, 123, 40, 37, 111, 52]. Recently, event-triggered control technique has been used successfully in various domains, like control of multi-agent system [32, 143, 62, 78, 83, 135], control over network [10, 122, 86], real-time embedded control [11, 85], etc. In event-triggered control, sensing, communication and, computation happen only when any predefined event-condition is violated. Generally, this event-condition is stated in terms of system's current states [117, 40, 57, 56] or outputs. To define a triggering condition and the closed loop stability for event-triggered system, most the references have exploited the ISS theory [115, 93, 68]. For ensuring the ISS property of such systems, a continuous-time Lyapunov function is considered which monotonically decreases at a predefined rate when input information is actuated at every aperiodic event-triggering instant. Later on, a discontinuous Lyapunov function is considered in [63], to state a co-design method of event-triggering condition and control law. Several researchers have extended the idea of event-triggered control for nonlinear systems [118, 141, 48, 64, 128]. The work in [118], solves an event-based trajectory tracking problem for a nonlinear system where the primary assumption is that all system states are measurable. To relax this assumption, an output feedback based event-triggered control law is proposed in [2, 1] for nonlinear system where the event-condition is expressed in terms of system output. Most of the references on event-triggered control consider a deterministic system model to derive the event-triggering condition. The research on event-triggered control for stochastic system [59, 33, 107, 142, 106] has been started by the pioneering work [12]. In [87], the stochastic event-triggered control of linear system is reported where the main objective of the work is to find a trade-off between closed-loop performance and communication resource allocation. In event-triggered control one of the prior assumption is that the processor is available at every event-triggering instant. To avoid such assumption, Gupta et al.[106] has proposed an event-triggered control algorithm of stochastic linear system for a time-varying processor availability. Along with the advantages of aperiodic sampling, proposing an event-triggered feedback control law for an uncertain system is a challenging research problem. The existing research works on event-triggered control for uncertain system are reviewed next.

Event-triggered control for uncertain system: In NCS, uncertainty is mainly considered in communicating network in the form of time-delay, data-packet loss etc. [102, 147]. On the other hand the unmodeled dynamics, time-varying system parameters, external disturbances are the primary sources of system uncertainties. The main shortcoming of the classical event-triggered system lies in the fact that one must know the exact model of the plant apriori. A system with an uncertain model is a more realistic scenario and has far greater significance. However, there are open problems in designing a control law and triggering conditions in the face of system uncertainties. Recently, several researchers have proposed event-based control algorithm for both uncertain linear and nonlinear systems.

To control a linear uncertain system through an event-triggered feedback law, various researchers [18, 30, 67] adopt sliding mode control (SMC) technique to regulate the system state. In [18], an event-based sliding mode control (SMC) technique has been reported where the state trajectories are attracted toward the sliding surface and stay within a sliding band afterward in spite of external disturbances. In the proposed control technique, the sliding band depends on system state and for a finite time it converges to its origin. Later on, the concepts of sliding mode event-triggered control is extended for uncertain nonlinear discrete-time system [30, 67]. Mostly in event-triggered control, the control input is applied only when the norm of mesurenment error $\|e\|$ crosses a constant threshold. To tackle unknown uncertiainity F. D. Brunner et al. [21] has proposed a discrete-time dynamic event-triggered control law where the value of threshold adapts dynamically based on uncertainty variation. The design of robust event-triggered control law based on system output is a challenging problem. In [34], V. S Dolk et al. have reported an novel output based event-triggered control law for linear continuous-time system. To regulate the system state, a dynamic controller and event-triggering mechanism have been considered. Apart from robust control method, several researchers [110, 30, 89] have proposed adaptive event-triggered control algorithms to regulate an uncertain system state. In [110], the adaptive control input has been generated using a NN implementation. The weight dynamics of NN has been written as an impulsive system. The weight vector of NN has been computed only when an event-condition is satisfied. In [126], an optimal adaptive even-triggered control law has been described for a nonlinear system. An actor-critic NN-based algorithm has been proposed to generate the control law. The Lyapunov based analysis has been used to prove the system stability. This monograph considers an optimal control approach for robust controller design to regulate the uncertain system behavior with limited feedback information. Next, a short review on optimal control approach for robust controller design is provided.

1.6.2 OPTIMAL CONTROL APPROACH FOR ROBUST CONTROLLER DESIGN

The design of robust feedback control law for an uncertain system is a very mature research area. Mainly, the unmodeled system dynamics, external disturbances are the primary sources of uncertainty. These uncertainties deteriorate the closed loop performance. To avoid such deterioration, a robust feedback control law is needed. In other words, the designed robust input can tolerate the uncertainties and maintains a satisfactory closed loop performance. In this monograph, the systems with parametric and model uncertainties are considered for analysis. The basic results of robust control method for linear uncertain system are reported in [16, 15, 24]. They have considered a linear time-varying uncertainty for analysis. In this subsection, the optimal control approach for robust control technique is reviewed.

Linear control law: F. Lin et al. [81, 79] have derived a novel robust control law for linear uncertain system using the results of optimal control theory. Both matched and mismatched uncertainties have been considered for design and analysis. To derive the control input, it has assumed that the uncertainty bound is known apriori. The upper bound of uncertainty is used to compute the control law. The essential idea of optimal control approach is an optimal control input computed for the nominal system, which minimizes a certain cost-functional. The derived optimal input for nominal system is the robust solution of original uncertain system. For linear system, the optimal control is computed by solving an infinite-time LQR problem. In LQR problem, an ARE is defined to generate the control input. To prove the system stability, a continuous time Lyapunov function is constructed using the solution of Riccati equation. The optimal control framework for robust control design has found application in tracking control of linear system [119]. In the work [119], initially the control law has been derived under the assumption that all states are measurable and then an investigation has been done for output feedback scenario. To generate the control inputs, a solution of a Riccati equation has been considered. Both analytical and simulation results have shown that the proposed control input enforces the system state to track the desired trajectory. The above introduced robust control law has been used to control a nonlinear systems [80, 82] where the system nonlinearity is treated as a source of uncertainty. To control such a nonlinear system, a linear robust control input is derived by solving a LQR problem for a linear nominal model and a cost-functional.

Nonlinear control law: For a nonlinear system, the robust control problem transforms into an equivalent nonlinear optimal control problem. The primary issue to solve a nonlinear optimal control problem is that the optimal input depends on the solution of HJB equation. Now, the computation of exact solution of HJB is difficult as it is a partial differential equation. Many researchers [76, 109, 103, 22] have used NN to approximate the solution of HJB equation as NN has universal function approximation property [53]. To solve the nonlinear optimal control problem using NN, the NN is trained through the open loop control inputs, which are computed at different points in state-space region. In [139], Wiener proposed another method to train the NN where a polynomial functions are considered to represent the system dynamics and input-output sequences. Recently, several researchers [19, 129, 136, 5] have adopted approximate dynamic programming (ADP) algorithm to compute the approximate solution of HJB equation. Based on an approximate solution of discrete-time HJB equation, the computation procedure of nonlinear optimal input have been reported in [5, 25, 113]. The approximate optimal control input has been computed using a greedy heuristic dynamic programming (HDP) method where the primary assumption is that the system model is perfectly known. But in practical scenario, the controller needs to be robust with respect to different uncertainties such as parametric variation of system model, disturbances, etc. In [24], a robust control law for nonlinear system has been proposed for matched system. It has shown that the proposed concepts can be applicable for both time-invariant and time-varying system. Recently, F. Lin et al. [80, 82] have used the solution of HJB equation to find out the robust control input. To solve the robust control problem, an equivalent optimal control problem has been formulated to derive the proposed robust control input. The optimal control problem is solved based on the nominal dynamics by minimizing a quadratic cost-functional with the knowledge of uncertainty bound. Similar concepts are used for nonlinear continuous-time system in [3, 4], where a non-quadratic cost-functional is considered. Recently D. Wang et.al. [132] have extended the F. Lin's approach [81, 79] for a discrete-time nonlinear system with matched uncertainty. But there are several physical systems like Magleve suspension system [73, 144], aircraft engine system [144], and the movement control of truck-trailer problem [148], where the so-called matching condition does not hold. Therefore considering mismatched uncertainty in system dynamics is a more realistic control problem.

1.7   SUMMARY

This chapter describes the basic idea of event-triggered control in the context of NCS. The event-triggered control technique is divided into two steps— controller design and event-triggering law design. The different triggering techniques and the procedure to derive the triggering laws have been discussed in this chapter. To design the triggering law, the Lyapunov and ISS theory have been used. The LQR-based controller design techniques has been introduced, to design the control law. The numerical examples are given to understand the different steps of event-triggered control technique. Both continuous and discrete-time systems are considered for design and analysis.

This chapter does not consider any model uncertainties in the system dynamics. But the presence of uncertainties in the system (1.4) affects the control law $u(t_k)$ and the triggering condition (1.46). In order to deal with system uncertainties with limited feedback information, a robust controller and an event-triggering rule need to be designed. A theoretical framework to design and analysis of robust event-triggered control law has been developed in the book and discussed in the subsequent chapters.

Section I

Control of Linear Systems

2 Robust Event-triggered Control for Continuous-time Linear Systems

After studying this chapter, one should be able to: define matched and mismatched uncertain systems; design robust controller for matched and mismatched systems using LQR method; model continuous-time event-triggered control system in the presences of uncertainties; design robust event-triggered controller and triggering law; define and derive dynamic event-triggering law; ensure the stability of event-triggered systems in the face of uncertainties; derive the lower bound of inter-event time; validate the control approach numerically.

2.1 INTRODUCTION

This chapter considers an optimal control framework to design a robust control law for an uncertain NCS with limited feedback information. The feedback information is limited due to bandwidth constraint of the communication channel. To capture the channel bandwidth limitations, an event-triggered control strategy is adopted in [117] without considering system uncertainty explicitly. With limited feedback information, existing robust control results in [81, 79] cannot be simply extended to the event-triggered system is the primary motivation for this work. This chapter proposes an event-based robust control strategy for both matched and mismatched uncertain systems. A conceptual block diagram of the proposed event-trigger-based robust control framework is illustrated in Fig. 2.1. Here the system, sensor, and actuator are co-located but the controller is connected through a communication network. A dedicated computing unit monitors the event condition at the sensor end. The control input is computed and actuated only when an event is generated. A zero-order-hold (ZOH) at the actuation end holds the last transmitted control input until the transmission of next input. The aperiodic state transmission to controller and control input update instant $\{t_k\}_{k \in \mathbb{N}}$ over the network is decided by the same event-triggering law. For simplification, it is assumed that there is no communication, computation, and actuation delay in the system. To design a robust control law for an uncertain system, an equivalent optimal control problem is formulated with

DOI: 10.1201/9781003229698-2

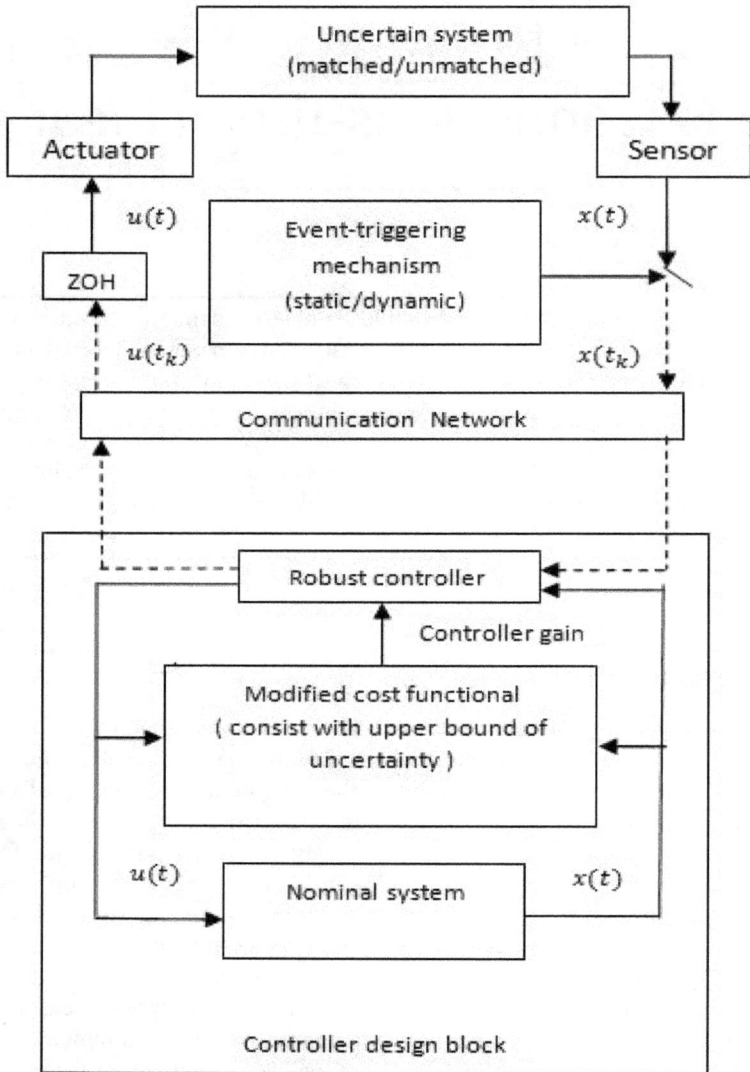

Figure 2.1 Conceptual block diagram of proposed event-trigger-based robust control. Dotted lines represent the aperiodic information transmission through the communication channel and $x(t_k)$ and $u(t_k)$ are state and control input, respectively.

an appropriate cost-functional, which takes care of the upper bound of system uncertainty. The nominal system dynamics is used to compute the optimal controller gain, which minimizes a cost-functional. The analysis of such uncertain system is done in continuous time domain. The proposed method is verified for both static and dynamic event-triggering rule. In both cases, corresponding triggering rule and their stability criteria for matched and mismatched uncertain system have been derived. The advantage of the proposed control strategy is that it significantly reduces the number of control input transmission and computation in spite of system uncertainties. The main contributions of this chapter are summarized as follows:

(i) Defining an optimal control problem to design a robust control law for both matched and mismatched uncertain systems.

(ii) Deriving static and dynamic event-triggering rule for a uncertain system using the upper bound of system uncertainty. A comparative study is carried out to verify the efficacy of both event-triggered robust control law with conventional approach.

(iii) Ensuring stability of closed loop system using an ISS Lyapunov function.

(iv) Deriving a positive non-zero lower bound of the inter-execution time.

Organization

The chapter is organized as follows. In Section 2.2, we briefly review an optimal approach to robust control design for the uncertain system. Sections 2.4 and 2.5 discuss the optimal control approach to solve the robust stabilization problem for the event-triggered uncertain system. Both static and dynamic event-triggering conditions are stated in the form of theorems and their corresponding proofs are reported. In Section 2.4 the expressions of the minimum positive inter-event time are defined. Two examples with simulation results are discussed in Section 2.6 to illustrate the proposed control algorithm. Section 2.7 summarizes the chapter.

2.2   ROBUST CONTROL

2.2.1   SYSTEM UNCERTAINTY

In general system, uncertainty is classified into two categories namely matched uncertainty and mismatched uncertainty. They are defined as follows:

System with matched uncertainty: A linear uncertain system is described by

$$\dot{x} = A(p)x + Bu \tag{2.1}$$

where $p \in \hat{P}$ is an uncertain parameter vector. The system (2.1) has matched uncertainty if there exists a bounded uncertain matrix $\phi(p) \in \mathbb{R}^{m \times n}$ such that

$$A(p) - A(p_0) = B\phi(p) \tag{2.2}$$

for any $p \in \hat{P}$, where p_0 is known nominal parameters and $A(p_0)$ is nominal system matrix. In other words system uncertainty is assumed to be in the range space of input matrix B. The condition (2.2) is made to simplify the derivation of stability results. It is assumed that there exists a positive semi-definite matrix F_m to represent the upper bound of the uncertainty i.e.,

$$\phi(p)^T \phi(p) \leq F_m, \; \forall \; p \in \hat{P}. \tag{2.3}$$

System with mismatched uncertainty: System (2.1) have mismatched uncertainty if its uncertainty is not in the range of input matrix B. In general system, uncertainty $(A(p) - A(p_0))$ can be decomposed in matched and mismatched components using pseudo-inverse B^+ of input matrix B [61]. Using $B^+ = (B^T B)^{-1} B^T$, the uncertainty in (2.1) can be written as

$$A(p) - A(p_0) \;\; = \;\; BB^+(A(p) - A(p_0)) + (I - BB^+)(A(p) - A(p_0)). \tag{2.4}$$

Here $BB^+(A(p) - A(p_0))$ is matched and $(I - BB^+)(A(p) - A(p_0))$ is an mismatched component of system uncertainty. It is assumed that $\forall \; p \in \hat{P}$ their exist $F_u \geq 0$ and $H \geq 0$, such that following holds:

$$(A(p) - A(p_0))^T (B^+)^T B^+ (A(p) - A(p_0)) \leq F_u, \tag{2.5}$$

$$\alpha^{-2}((A(p) - A(p_0))^T (A(p) - A(p_0)) \leq H. \tag{2.6}$$

Here the scalar $\alpha \geq 0$ is a design parameter. To stabilize (2.1) with matched uncertainty (2.2) (or mismatched uncertainty (2.4)), we need to design a robust controller, which is discussed next.

2.2.2   ROBUST CONTROL PROBLEM

Find a static state feedback control law $u = Kx$ such that the uncertain system (2.1) is stable with (2.2) or (2.4) for any $p \in \hat{P}$.

To solve the above mentioned robust control problem, this chapter has adopted an optimal control approach. The essential idea is to compute the optimal control input for the nominal system which minimizes a modified cost-functional. Then, the optimal control input for nominal system is shown to be a robust control input for the actual uncertain system. Here the system (2.1) may have matched or mismatched uncertainty. In both cases their corresponding nominal dynamics and cost-functional are discussed as follows.

Nominal dynamics and cost-functional for matched uncertain system are described as

$$\begin{cases} \dot{x} = A(p_0)x + Bu_1 \\ u_1 = K_1 x \end{cases} \tag{2.7}$$

$$J_m = \int_0^\infty (x^T F_m x + x^T Q x + u_1^T R u_1) dt \tag{2.8}$$

with $Q \geq 0$ and $R > 0$. The matrix $F_m \geq 0$ is the upper bound of matched uncertainty and it is defined in (2.3).

Auxiliary dynamics and cost-functional for the mismatched uncertain system are defined as

$$\begin{cases} \dot{x} = A(p_0)x + Bu_2 + \alpha(I - BB^+)v \\ u_2 = K_2 x \\ v = Lx \end{cases} \tag{2.9}$$

$$J_u \quad = \quad \int_0^\infty (x^T(F_u + \rho^2 H + \beta^2 I)x + u_2^T u_2 + \rho^2 v^T v)dt. \tag{2.10}$$

The auxiliary input v will be used to derive existence condition for the robust stability of mismatched uncertain system. Now to design a robust control law using optimal control approach, following lemma is introduced [81, 79, 3].

Lemma 3 *Suppose there exists an optimal control solution of nominal system (2.7) for matched system [(2.9) for mismatched system] with a cost-functional (2.8) [(2.10) for mismatched system]. Then the optimal control law for the nominal system will be the robust control solution of the original system (2.1) for all bounded system uncertainty (2.2) [mismatched uncertainty (2.4)].*

Proof of Lemma 3 *The present approach translates the robust control problem stated in Section 2.2 to an equivalent optimal control problem [81, 4]. The proof consists of two parts.*
Stability proof for matched uncertain system: *If $V(x)$ is the Lyapunov function for (2.7), then the time derivative of $V(x)$ along the state trajectory of (2.1) is*

$$\dot{V}(x) \quad = \quad V_x \dot{x} = V_x^T(A(p_0)x + BK_1 x) + V_x^T B\phi(p)x.$$

Using (2.20), (2.22) and by adding and subtracting $x^T\phi(p)^T R\phi(p)x$, the $\dot{V}(x)$ reduces to

$$\begin{aligned} \dot{V}(x) \quad &= \quad -x^T((F_m - \phi(p)^T R\phi(p)) + Q + (K_1 + \phi(p))^T R(K_1 + \phi(p))x \\ &\leq \quad -x^T x. \end{aligned} \tag{2.11}$$

Now $\dot{V}(x) \leq 0$ for $x \neq 0$, which ensures asymptotic stability of the original uncertain matched systems (2.1) and (2.2).
Stability proof for unmatched uncertain system: *Similarly for a Lyapunov function $V(x)$, the $\frac{dV(x)}{dt}$ along the state trajectory of (2.1) is simplified using (2.5)–(2.6) and (2.26)–(2.28) as*

$$\begin{aligned} \dot{V}(x) \quad &\leq \quad -x^T(F_u + \rho^2 H + \beta^2 I)x - \rho^2 x^T L^T L x + x^T F_u x \\ &\quad + \rho^2 x^T L^T L x + \rho^2 x^T H x + 2\rho^2 x^T L^T L x \\ &= \quad -x^T(\beta^2 I - 2\rho^2 L^T L)x. \end{aligned} \tag{2.12}$$

Now $\dot{V}(x) \leq 0$ for $x \neq 0$ if the inequality $\beta^2 I - 2\rho^2 L^T L > 0$ holds. Therefore the closed loop system (2.1), (2.4) is asymptotically stable for all $p \in P$. From the above two proofs, the optimal control input u_1 of (2.7) or u_2 of (2.9) are the robust control input for (2.1), which proves the Lemma 3.

2.3 PROBLEM DESCRIPTION AND STATEMENT

In this section, the aperiodic feedback information of system states are used to solve the robust control problem. This formulation helps to realize such a controller in a networked control domain with limited state information. The aperiodic control input computation instant is determined through a state-dependent event condition, which is derived from the stability results. Suppose $\{t_k\}$ represents (aperiodic state transmission, control input computation, and actuation instant) the event occurring instant. The event-based state feedback control input will be

$$u(t_k) = Kx(t_k) \tag{2.13}$$

which replaces the usual continuous time state feedback control law $u(t) = Kx(t)$. To solve the robust control problem through an aperiodic control law (2.13) the uncertain linear system (2.1) is rewritten as

$$\dot{x} = A(p)x(t) + Bu(t_k). \tag{2.14}$$

Adopting the concepts introduced in [117], the event-based closed loop system (2.14) reduces to

$$\dot{x} = A(p)x + BK(x + e). \tag{2.15}$$

The variable $e \in R^n$ is referred as measurement error and defined as

$$e(t) = x(t_k) - x(t), \forall t \in [t_k, t_{k+1}), k \in \mathbb{N}. \tag{2.16}$$

Using (2.15), (2.7), and (2.2) the event-triggered system with matched uncertainty is described as

$$\dot{x} = A(p_0)x + Bu_1 + B(\phi(p)x + K_1 e). \tag{2.17}$$

Similarly for mismatched uncertainty (2.4), the event-triggered system (2.15) is

$$\begin{aligned}
\dot{x} &= A(p_0)x + BK_2(x + e) + \alpha(I - BB^+)L(x + e) + BB^+(A(p) - A(p_0))x \\
&\quad + (I - BB^+)(A(p) - A(p_0))x - \alpha(I - BB^+)L(x + e).
\end{aligned} \tag{2.18}$$

Problem statement: Design the controller gain K_1 for (2.17) and K_2 along with L for (2.18) to stabilize an uncertain event-triggered system (2.17) and (2.18) such that the closed loop system is ISS with respect to its measurement error e [defined in (2.16)]. We attempt to solve these problems in two steps. Firstly, design a controller using Lemma 3 and then define an event-triggering rule such that the closed loop system (2.17) [or (2.18)] is ISS. These two solution steps are discussed next.

2.3.1 CONTROLLER DESIGN

The system (2.7) is the nominal dynamics of (2.17) for matched system. Using Lemma 3, the optimal controller gain K_1 of (2.7) which minimizes the cost-functional (2.8) will be the robust solution of (2.17). Similarly for mismatched system, gains K_2 and L of (2.9) that minimizes (2.10), is the robust solution for (2.18).

Step 1 For matched system, control input $u_1(t)$ is designed by minimizing J_m. Suppose $V(x)$ be a Lyapunov function for (2.7). Now using optimal control results, $V(x)$ should satisfy the Hamilton-Jacobi-Bellman (HJB) equation [91, 79]

$$min_{u_1 \in R^m}(x^T F_m x + x^T Q x + u_1^T R u_1 + V_x^T (A(p_0)x + B u_1)) = 0 \qquad (2.19)$$

where $V_x = \frac{\partial V}{\partial x}$ and $u_1 = K_1 x$. Applying optimal u_1, (2.19) reduces to

$$(x^T F_m x + x^T Q x + u_1^T R u_1 + V_x^T (A(p_0)x + B u_1)) = 0. \qquad (2.20)$$

Step 2 According to optimal control theory, the optimal input $u_1(t)$ should minimize the Hamiltonian [91]

$$H(x(t), u_1(t), V_x) = x^T F x + x^T Q x + u_1^T R u_1 + V_x^T (A(p_0)x + B u_1) \qquad (2.21)$$

which leads to

$$\frac{\partial H(x(t), u_1(t), V_x, t)}{\partial u_1(t)} = 2x^T K_1^T R + V_x^T B = 0. \qquad (2.22)$$

Step 3 For solving an infinite-time LQR problem, a quadratic function $V(x) = x^T S x$ is defined, where matrix $S > 0$. With this choice, the HJB equation (2.20) reduces to the following ARE

$$SA(p_0) + A(p_0)^T S + F_m + Q - SBR^{-1}B^T S = 0. \qquad (2.23)$$

Using the solution S of (2.23), the optimal control input u_1 is computed as

$$u_1(t) = -R^{-1}B^T S x(t) = K_1 x(t). \qquad (2.24)$$

Step 4 Control gain K_1 for (2.7) and aperiodic state information of original system $x(t_k)$ are used to realize the event-triggered control law

$$u_1(t_k) = K_1 x(t_k). \qquad (2.25)$$

The above mentioned steps 1 to 4 are also adopted for the mismatched system (2.9) with cost-functional J_u (2.10). The control input $u_2(t)$ and auxiliary input $v(t)$ are computed if the following equations are satisfied:

$$x^T (F_u + \rho^2 H + \beta^2 I)x + u_2^T u_2 + \rho^2 v^T v + V_x^T (A(p_0)x$$
$$+ B u_2 + \alpha(I - BB^+)v) = 0, \qquad (2.26)$$

$$2x^T K_2^T + V_x^T B = 0, \qquad (2.27)$$

$$2\rho^2 x^T L^T + V_x^T \alpha(I - BB^+) = 0. \qquad (2.28)$$

For solving infinite-time LQR problem, select a Lyapunov function as $V(x) = x^T \hat{S} x$. The HJB equation (2.26) reduces to the following ARE

$$\hat{S}A(p_0) + A(p_0)^T \hat{S} + F_u + \rho^2 H + \beta^2 I - \hat{S}(BB^T + \alpha^2 \rho^{-2}(I - BB^+)(I - BB^+)^T)\hat{S} = 0. \quad (2.29)$$

Using the solution $\hat{S} > 0$ of (2.29), control input u_2 and auxiliary input v are computed as

$$\begin{bmatrix} u_2(t) \\ v(t) \end{bmatrix} = \begin{bmatrix} -B^T \hat{S} \\ -\alpha \rho^2 (I - BB^+)\hat{S} \end{bmatrix} = \begin{bmatrix} K_2 \\ L \end{bmatrix} x(t). \tag{2.30}$$

The robust event-triggered control input for (2.18) is written as

$$u_2(t_k) = K_2 x(t_k). \tag{2.31}$$

Now for event-triggered control, it is important to design the event-triggering instant t_k such that uncertain system (2.14) is ISS with an aperiodic control law (2.13). The approach for deriving the triggering law is discussed next.

2.3.2   TRIGGERING CONDITION DESIGN

Given an uncertain system (2.14) with a linear controller (2.13), there should be an event-triggering instant $t_{k \in \mathbb{N}}$ with a positive inter-execution time $(t_{k+1} - t_k = \tau > 0)$ such that the closed loop system (2.14) is ISS. To prove this, an ISS Lyapunov function is considered such that its time derivative is in the form of (1.35). The ISS condition in the form of (1.35) helps to construct the event-triggering rule in-terms of measurement error norm $\|e(t)\|$ and the state norm $\|x(t)\|$. To design the event-triggering law for matched system (2.17) and mismatched system (2.18), the ISS Lyapunov functions are considered as follows:

ISS Lyapunov function for matched system:

$$V_m(x) = x^T S x \tag{2.32}$$

ISS Lyapunov function for mismatched system:

$$V_u(x) = x^T \hat{S} x \tag{2.33}$$

2.4   STATIC EVENT-TRIGGERED ROBUST CONTROL

This section describes static event-triggering law for both matched and mismatched uncertain systems. The main results of this chapter are stated in the form of following theorems.

Theorem 1 *Suppose the controller gain matrix K_1 is designed for the nominal system (2.7) by minimizing the cost-functional (2.8). The matched uncertain system (2.17), with even-triggered controller (2.25), is ISS if there exists a static event occurring sequence $\{t_k\}_{k \in \mathbb{N}}$ given by*

$$t_0 = 0, \ t_{k+1} = \inf\{t \in \mathbb{R} | t > t_k \wedge \mu_1 \|x\| - \|e\| \leq 0\} \tag{2.34}$$

where parameter μ_1 is defined in (2.40).

Proof *To prove ISS of (2.17), it is necessary to simplify the derivative of ISS Lyapunov function $V(x)$ in the form of (1.35). Here the expression of $V(x)$ is same as $V_m(x)$, as defined in (2.32). The time derivative of $V(x)$ along the solution of (2.17) can be written as*

$$\dot{V}(x) = V_x^T \dot{x} \tag{2.35}$$
$$= V_x^T (A(p_0)x + BK_1x + B\phi(p)x + BK_1e).$$

Using (2.20) and substituting $V_x^T = 2x^T S$, following simplifications are made

$$\dot{V}(x) = -x^T F_m x - x^T Q x - u^T R u - 2x^T K_1^T R\phi(p)x + 2x^T SBK_1 e$$
$$= -x^T \{(F_m - \phi(p)^T R\phi(p)) + Q + (K_1 + \phi(p))^T R(K_1 + \phi(p))\}x$$
$$+ 2x^T SBK_1 e.$$

According to Definition 8, $\dot{V}(x)$ can be written in the form (1.35) if we select

$$\alpha(\|x\|) = \frac{\lambda_{min}(Q_1)}{2}\|x\|^2, \tag{2.36}$$

$$\gamma(\|e\|) = \frac{2\|SBK_1 K_1^T B^T S\|}{\lambda_{min}(Q_1)}\|e\|^2. \tag{2.37}$$

In the above expressions, the matrix Q_1 is

$$Q_1 = (F_m - \phi(p)^T R\phi(p)) + Q + (K_1 + \phi(p))^T R(K_1 + \phi(p)). \tag{2.38}$$

From (1.35), (2.36), and (2.37), triggering law is derived as

$$\|e\| \leq \mu_1\|x\| \tag{2.39}$$

where

$$\mu_1 = \frac{\sigma^{1/2}\lambda_{min}(Q_1)}{2\|SBK_1\|}. \tag{2.40}$$

Note that the condition (2.39) needs to be violated to update the control input. Moreover the time instant at which the event has occurred is given by (2.34). Using (2.39), $\dot{V}(x)$ becomes

$$\dot{V}(x) \leq (\sigma - 1)\lambda_{min}(Q_1)\|x\|^2.$$

Therefore, the event-triggered system (2.17) is ISS $\forall \, \sigma \in (0, 1)$.

Remark 4 *It is seen from (2.3) that the first term of Q_1 is a positive definite matrix. The bound on final term of Q_1 is derived as $(K_1 + \phi(p))^T R(K_1 + \phi(p)) \leqslant \|K_1^T RK_1\| + \lambda_{max}(RF_m)$. The positiveness of all three terms ensures the positive definiteness of Q_1.*

Remark 5 *In the absence of uncertainty, $\phi(p) = 0$, the expression (2.34) reduces to the similar results reported in [117, 52]. In this sense, the proposed algorithm generalizes the existing results and also valid for uncertain systems.*

The results for mismatched system is stated in the form of following theorem:

Theorem 2 *Suppose the controller gain matrices K_2 and L are designed for nominal system (2.9) by minimizing the cost-functional (2.10) and the inequality $\beta^2 I - 2\rho^2 L^T L > 0$ holds. The uncertain system (2.18), with event-based control law (2.31) is ISS if there exists a static event occurring sequence $\{t_k\}_{k \in N}$ given by*

$$t_0 = 0, t_{k+1} = inf\{t \in \mathbb{R} | t > t_k \wedge \mu_2 \|x\| - \|e\| \le 0\} \tag{2.41}$$

where the design parameter μ_2 is defined in (2.47).

Proof *Assume $V_u(x)$ is an ISS Lyapunov function of (2.18) and denote $V_u(x)$ by $V(x)$. The time derivative of $V(x)$ along the state-trajectory of (2.18) is simplified as*

$$\begin{aligned}
\dot{V}(x) &= V_x^T \dot{x} \\
&= V_x^T (A(p_0)x + BK_2 x + BK_2 e) + V_x^T \alpha(I - BB^+)Lx + V_x^T \alpha(I - BB^+)Le \\
&\quad + V_x^T BB^+ (A(p) - A(p_0))x + V_x^T (I - BB^+)(A(p) - A(p_0))x \\
&\quad + V_x^T \alpha(I - BB^+)(Le + Lx).
\end{aligned}$$

Using (2.26), (2.27), and (2.28)

$$\begin{aligned}
\dot{V}(x) &= -x^T \{(F_u + \rho^2 + \beta^2 I) + K_2^T K_2 + \rho^2 L^T L + 2K_2^T B^T (A(p) - A(p_0)) \\
&\quad + 2\alpha^{-1} \rho^2 L^T (A(p) - A(p_0))\}x + 2x^T \tilde{S}BK_2 e.
\end{aligned}$$

Using the upper bound in (2.5), (2.6), the above equality is simplified as

$$\dot{V}(x) \le -\lambda_{min}(Q_2) \|x\|^2 + \|\tilde{S}BK_2 K_2^T B^T \tilde{S}\| \|e\| \|x\|. \tag{2.42}$$

Now as per Definition 8, the condition (1.35) holds if

$$\alpha(\|x\|) = \frac{\lambda_{min}(Q_2)}{2} \|x\|^2, \tag{2.43}$$

$$\gamma(\|e\|) = \frac{2\|\tilde{S}BK_2 K_2^T B^T \tilde{S}\|}{\lambda_{min}(Q_2)} \|e\|^2. \tag{2.44}$$

In the above expression (2.43), the matrix Q_2 is

$$Q_2 = \beta^2 I - 2\rho^2 L^T L. \tag{2.45}$$

By hypothesis of Theorem 2, the matrix Q_2 is positive-definite. Using (1.35), (2.43), and (2.44) it can be concluded that control input should be updated if the following triggering condition is violated:

$$\|e\| \le \mu_2 \|x\| \tag{2.46}$$

where

$$\mu_2 = \frac{\sigma^{1/2}\lambda_{min}(Q_2)}{2\|\tilde{S}BK_2\|}. \tag{2.47}$$

Equations (2.46), (2.47) also suggest the time instant when the condition (2.46) does not hold and it is expressed as (2.41). Using (2.46), $\dot{V}(x)$ reduces to

$$\dot{V}(x) \leqslant (\sigma - 1)\lambda_{min}(Q_2)\|x\|^2 \tag{2.48}$$

which ensure the negative definiteness of $\dot{V}(x)$, $\forall\, \sigma \in (0,1)$.

Theorems 1 and 2 ensure stability of uncertain systems (2.17) and (2.18) by static event-triggering rule, respectively. An algorithmic representation of static event-triggered control with matched system uncertainty is given next.

Algorithm 1 Static event-triggered control for matched uncertain system

---

1: Initialization: $t \Leftarrow 0$, $x \Leftarrow x_0$, $x(t_k) \Leftarrow x_0$.
2: Given: $A(p_0)$, B, F_m, σ
3: Compute μ_1 using (2.40)
4: **if** $\|e\| \geq \mu_1\|x\|$ **then**
5:     Transmit $x(t_k)$ from system to controller.
6:     Solve (2.23) and compute K_1.
7:     Compute and update $u_1(t_k) = K_1x(t_k)$ in (2.25).
8: **else**
9:     $u_1(t) = u_1(t_{k-1})$
10: **end if**
11: Return to line 3

Minimum time interval in between two consecutive events

In event-triggered control, the inter execution time depends on the evolution of $\|e\|/\|x\|$ with respect to time. At t_k the ratio of $\|e\|/\|x\|$ is zero as measurement error $e = 0$. The next event will occur at t_{k+1}, when the ratio $\|e\|/\|x\|$ turns to μ_1. Using (2.41), the minimum time required to evolve $\|e\|/\|x\|$ from 0 to μ_1 defines the lower bound of the inter-event time $\{t_{k+1} - t_k\}_{\forall k \in \mathbb{I}} = \tau > 0$. Here inter-event time τ should be always a non-zero positive time interval to avoid the so-called Zeno behavior. The minimum time interval between two consecutive events of proposed robust control mechanism is stated in the form of a theorem.

Theorem 3 $\forall\, \sigma \in (0,1)$, *the system (2.17) with triggering law (2.34) has strictly positive lower bound of inter-event time $\tau > 0$ and it is expressed as*

$$\tau = \frac{2}{\sqrt{L_3^2 - 4L_2L_1}}\ln\left\|\frac{(N_1+N_2)N_3}{(N_1+N_3)N_2}\right\| \tag{2.49}$$

where $N_1 = 2L_2\mu_1$, $N_2 = L_3 - \sqrt{L_3^2 - 4L_2L_1}$, $N_3 = L_3 + \sqrt{L_3^2 - 4L_2L_1}$ and $L_1 = \|(A(p_0) + B\phi(p) + BK)\|$, $L_2 = \|BK\|$, $L_3 = L_1 + L_2$.

Proof *From [117], the time derivative of $\|e\|/\|x\|$ can be written as*

$$\frac{d}{dt}\frac{\|e\|}{\|x\|} \leq \frac{\|e\|\|\dot{x}\|}{\|e\|\|x\|} + \frac{\|x\|\|\dot{x}\|}{\|x\|\|x\|}\frac{\|e\|}{\|x\|}$$

$$\leq \left(1 + \frac{\|e\|}{\|x\|}\right)\frac{\|\dot{x}\|}{\|x\|}. \tag{2.50}$$

Applying triangular inequality of vector norm on (2.17), the following simplification is made

$$\|\dot{x}(t)\| \leq \|(A(p_0) + B\phi(p) + BK_1)\|\|x\| + \|BK_1\|\|e\|. \tag{2.51}$$

Denoting $L_1 = \|(A(p_0) + B\phi(p) + BK_1)\|$, $L_2 = \|BK_1\|$ and the ratio $\frac{\|e\|}{\|x\|} = y$, the inequality (2.50) is simplified as

$$\frac{dy}{dt} \leq L_1 + (L_1 + L_2)y + L_2 y^2. \tag{2.52}$$

Applying Comparison lemma[1] [71] on (2.52), the differential inequality (2.52) turns into the following equality

$$\dot{\Omega} = L_1 + (L_1 + L_2)\Omega + L_2\Omega^2 \tag{2.53}$$

With a initial value $\Omega(0, \Omega_0) = \Omega_0$, the solution $\Omega(t, \Omega_0)$ of (2.53) satisfies the inequality $y(t) \leq \Omega(t, \Omega_0)$. Thus the inter-event time, τ is bounded by a finite amount of time to evolve Ω from 0 to μ_1. The expression of τ can be derived by solving (2.53) and given as

$$\tau = \frac{2}{\sqrt{L_3^2 - 4L_2L_1}}\ln\left\|\frac{(N_1 + N_2)N_3}{(N_1 + N_3)N_2}\right\|. \tag{2.54}$$

From (2.54) it is obvious that τ has positive value as $N_3 > N_2$.

Remark 6 *For mismatched system, expression of τ is similar to (2.49) but the value of L_1, L_2, and L_3 are $L_1 = \|A(p_0) + BB^+(A(p) - A(p_0)) + (I - BB^+)(A(p) - A(p_0)) + BK\|$, $L_2 = \|BK\|$, and $L_3 = L_1 + L_2$. In both cases, the expressions of L_1 and L_3 depend on system uncertainty. Therefore to compute the lower bound of inter-event time, the value of L_1 and L_3 are computed in entire uncertainty region such that τ is minimal. It is possible as the bound on uncertainty is known for both matched and mismatched systems.*

---

[1]The basic concept of comparison lemma has been stated in Appendix A.

2.5 DYNAMIC EVENT-TRIGGERED ROBUST CONTROL

A. Girad proposed a dynamic event-triggering mechanism where a dynamic variable $\eta(t) \geq 0$ is added to achieve larger inter-event time [52]. The evolution of new variable $\eta(t)$ with respect to time is expressed by the following general differential equation.

$$\dot{\eta}(t) = -\beta(\eta(t)) + \sigma\alpha(\|x(t)\|) - \gamma(\|e(t)\|). \tag{2.55}$$

Here β, α, γ are smooth class $\mathscr{K}_\infty$ functions and $\sigma \in (0,1)$. The preliminaries and efficiency of dynamic event-triggering mechanism over the static one [117] is reported in [52]. In this section, the dynamic event-triggering approach is adopted to solve the robust control problem with limited state information. Equations (2.17) and (2.55) are used to define an augmented uncertain system as

$$\begin{cases} \dot{x} = A(p_0)x + Bu_1 + B(\phi(p)x + K_1 e) \\ \dot{\eta}(t) = -\beta(\eta(t)) + \sigma\alpha(\|x(t)\|) - \gamma(\|e(t)\|). \end{cases} \tag{2.56}$$

The augmented system (2.56) helps to derive the event-triggering rule. The dynamic event-triggering instant generated for (2.56) is stated in form of a theorem.

Theorem 4 *Suppose the controller gain matrix K_1 is designed for the nominal system (2.7) by minimizing the cost-functional (2.8). The augmented matched system (2.56) with event-trigger-based controller (2.25) is ISS if there exists an event occurring sequence $\{t_k\}_{k\in\mathbb{I}}$ given by*

$$t_0 = 0,\ t_{k+1} \quad = \quad \inf\{t \in \mathbb{R} | t > t_k \wedge \eta(t) + \theta(\mu_1\|x\| - \|e\|) \leq 0\} \tag{2.57}$$

where μ_1 is defined in (2.40).

Proof *From (2.55), the evolution of $\eta(t)$ with respect to time can be defined as*

$$\dot{\eta}(t) = -\lambda\eta(t) + (\mu_1\|x\| - \|e\|),\ \eta(0) > 0. \tag{2.58}$$

Now select $W(x(t), \eta(t)) = V_m(x) + \eta(t)$ as a Lyapunov function for the following augmented system

$$\begin{cases} \dot{x} = A(p_0)x + Bu_1 + B(\phi(p)x + K_1 e) \\ \dot{\eta}(t) = -\lambda\eta(t) + (\mu_1\|x\| - \|e\|),\ \eta(0) > 0 \end{cases} \tag{2.59}$$

where $V_m(x)$ is defined in (2.32). Using (2.35) and (2.58) the time derivative of $W(x)$ can be written as

$$\dot{W}(x) \leq (\sigma - 1)\lambda_{min}(Q_1)\|x\|^2 - \lambda\eta(t). \tag{2.60}$$

From (2.60), for any value of $\sigma \in (0,1)$ and $\eta(t) > 0$, the closed loop system (2.17) is ISS by (2.57).

> **Remark 7** *Dynamic event-triggering law for mismatched system is not discussed here as the derivation will be similar in nature. In that case triggering law will depend on parameter μ_2 and Q_2, defined in (2.47), (2.45). The stability proof will be similar to the proof of Theorem 4.*

The expression of inter-event time τ for dynamic event-triggered case is derived next.

2.5.1 SELECTION OF DESIGN PARAMETERS

The parameters θ, σ, and λ are used in (2.57)–(2.60). These parameters mainly affect the lower bound of inter-event time and convergence rate of system state. This subsection discusses a possible methodology for selection of such parameters. The convergence of closed loop system (2.17) and (2.18) is directly associated with σ as seen in (2.60). As $\sigma \to 0$, the convergence rate of (2.17) [or (2.18)] is equivalent to the ideal closed loop system (2.1). The generated event number can also be controlled by varying the value of σ. Similarly the parameter θ has contribution in determining the inter-event time τ. A possible selection procedure of parameter θ is carried out by deriving a lower bound on τ. The results are stated in the form of a theorem.

> **Theorem 5** $\forall\ \sigma \in (0,1),\ \eta > 0$ *and* $\theta > 0$ *the system (2.17), (2.58) with triggering law (2.57) has strictly positive lower bound of inter-event time* $\tau > 0$ *and it is expressed as*
>
> $$\tau = \int_0^{\mu_1} \frac{d\Gamma}{\frac{L_1}{\mu_1} + (L_2 + \lambda)\Gamma + (\frac{1}{\theta} + L_2\mu_1)\Gamma^2} \tag{2.61}$$
>
> *where* $L_1 = \|(A(p_0) + B\phi(p) + BK)\|$, $L_2 = \|BK\|$ *and* $0 < \theta \le \frac{1}{L_1 - \lambda}$.

Proof of Theorem 5 *In dynamic event-triggered control the inter-event time depends on the evolution of* Γ, *where*

$$\Gamma = \frac{\theta \|e\|}{\eta + \theta \mu_1 \|x\|} \tag{2.62}$$

The time derivative of Γ *along the direction of state trajectories in (2.17), (2.58) is simplified as*

$$\dot{\Gamma} \le \frac{\theta(L_1 \|x\| + L_2 \|e\|)}{(\eta + \theta \mu_1 \|x\|)} + \frac{\theta \|e\|}{(\eta + \theta + \mu_1 \|x\|)^2}$$
$$\times \left\{ \lambda \eta - \mu_1 \|x\| + \|e\| + \theta \mu_1 L_1 \|x\| + \theta \mu_1 L_2 \|e\| \right\}$$
$$\le \frac{L_1}{\mu_1} + (L_2 + \lambda)\Gamma + (\frac{1}{\theta} + L_2\mu_1)\Gamma^2 + \frac{\theta \mu_1 \|x\|}{(\eta + \theta \mu_1 \|x\|)} \left(-\lambda - \frac{1}{\theta} + L_1 \right)\Gamma. \tag{2.63}$$

Selecting $\theta = \frac{1}{L_1 - \lambda}$, *the final term in (2.63) reduces to zero. Adopting the similar steps as described in Section 2.4, the lower bound of inter-event time for dynamic event-triggering is*

$$\tau = \int_0^{\mu_1} \frac{d\Gamma}{\frac{L_1}{\mu_1} + (L_2 + \lambda)\Gamma + (\frac{1}{\theta} + L_2\mu_1)\Gamma^2}. \tag{2.64}$$

To prove the positiveness of τ *consider the function*

$$g(\Gamma) = \frac{L_1}{\mu_1} + (L_2 + \lambda)\Gamma + (\frac{1}{\theta} + L_2\mu_1)\Gamma^2 \tag{2.65}$$

which has all positive coefficients. The function (2.65) is a positive function as $\frac{dg}{d\Gamma} > 0, \ \forall \Gamma > 0$. *Note that,* Γ *is a positive variable as* η, θ, *and* μ_1 *all are positive. Therefore integration of (2.65) over any positive interval is always positive. The expression (2.64) is also valid for* $0 < \theta < \frac{1}{L_1 - \lambda}$. *This completes the proof.*

Remark 8 *The existence of positive inter-event time is guaranteed in the range of* $0 < \theta \leq \frac{1}{L_1 - \lambda}$ *and it helps to select the other parameter* λ. *The value of* λ *must satisfy* $\lambda \leq L_1$ *in order to make* θ *positive.*

Remark 9 *The expression of* τ *in (2.61) is derived for* $0 < \theta \leq \frac{1}{L_1 - \lambda}$. *Similarly, an analytical bound on* τ *can also be derived for* $\theta > \frac{1}{L_1 - \lambda}$. *Note that the value of scalar* L_1 *depends on uncertainty* $\phi(p)$. *Hence, it is difficult to say the exact value of* θ *for which event-triggering law (2.57) has larger lower-bound* τ. *But it is possible to compute* τ *as the uncertain region is known apriori.*

Remark 10 *The analytical expression of* τ *for mismatched system is not addressed here. But it can be derived using similar approach with different* $L_1, L_2,$ *and* L_3. *The existence of larger average inter-event time of dynamic event-triggering rule over the static one is shown numerically in the next section.*

2.6 SIMULATION RESULTS AND COMPARISONS

This section explains two separate numerical examples to validate the theoretical results for both matched and mismatched event-triggered systems.

2.6.1 EXAMPLE 1

A second-order linear system with matched uncertainty is shown below:

$$\dot{x} = A(p)x + Bu_1$$

where $A(p) = \begin{bmatrix} 0 & 1 \\ 1+p & p \end{bmatrix}$, $B = \begin{bmatrix} 0 \\ 1 \end{bmatrix}$ and the uncertain vector $p \in [-2,2]$. To solve (2.23), different matrices are selected as $Q = \begin{bmatrix} 10 & 0 \\ 0 & 10 \end{bmatrix}$, $R = 2$, and $F_m = \begin{bmatrix} 8 & 8 \\ 8 & 8 \end{bmatrix}$.

Using these matrices, the solution of (2.23) is obtained as $S = \begin{bmatrix} 16.89 & 6.89 \\ 6.89 & 6.89 \end{bmatrix}$. The controller gain K_1 is calculated as $K_1 = \begin{bmatrix} -4.162 & -4.162 \end{bmatrix}$. The scalar $\lambda_{min}(Q_1) = 10$ is calculated from (2.38) using maximum upper bound of the uncertain parameter $p = 2$. To compute t_k, the design parameters are selected as $\sigma = 0.98$, $\theta = 0.1$, and $k = 0.6$. The simulation is executed for 4.5 second with the initial states $[0.2, -0.35]^T$ for static and $[0.2, -0.35, 0.01]^T$ for dynamic event-triggered control. Here the parameter p varies sinusoidally according to the equation $p = 2sin(t)$. It can be seen in Figs. 2.2(b) and 2.3(b) that the error norm is bounded by a state dependent threshold. This signifies that the closed loop system holds the ISS property. Figures 2.2(a) and 2.3(a) show the total number of events and their corresponding positive inter-execution time for a matched system. From Table 2.1,

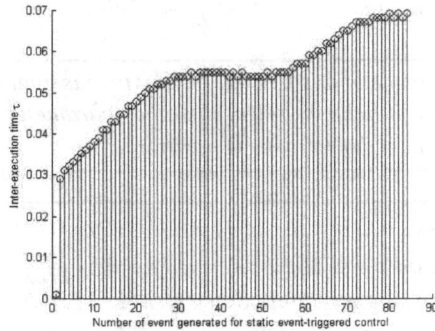

(a) Number of event occurrence and time interval between two consecutive events for static event-triggered control.

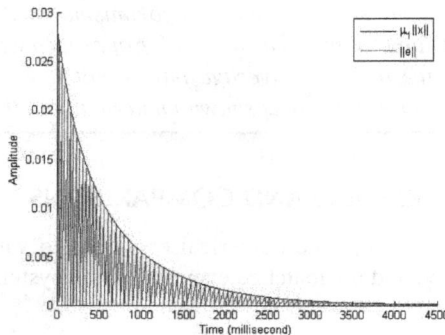

(b) Time evolution of $\|e\|$ (which is always within the threshold limit $\mu_1 \|x\|$) for static event-triggered control.

Figure 2.2 Results of static event-triggered control with matched uncertainty.

Table 2.1

Comparative Results of Event-triggered and Conventional Continuous Robust Control Approach. Subscripts "M" and "U" Stands for Matched and Mismatched System, Respectively.

Control mechanism	τ_{max}	τ_{min}	τ_{avg}	u_{total}
Without event-triggered control	0.001	0.001	0.001	3001
Static event-triggered control	0.06_M 0.14_U	0.001_M 0.01_U	0.05_M 0.07_U	86_M 43_U
Dynamic event-triggered control	0.10_M 0.44_U	0.06_M 0.05_U	0.08_M 0.16_U	55_M 19_U

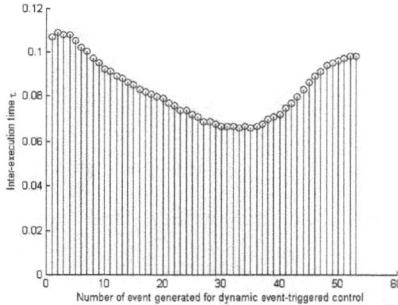

(a) Number of event occurrence and time interval between two consecutive events for dynamic event-triggered control.

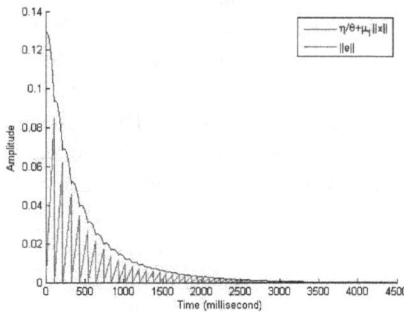

(b) Time evolution of $\|e\|$ (which is always within the threshold limit $\{\eta/\theta + \mu_1\|x\|\}$) for dynamic event-triggered control.

Figure 2.3 Results of dynamic event-triggered control with matched uncertainty.

it is seen that the average inter-actuation time for event-based control is significantly larger than the conventional continuous one.

2.6.2 EXAMPLE 2

We consider a second-order mismatched uncertain system (2.18) where $A(p) = \begin{bmatrix} 0 & 1+p \\ 1 & 0 \end{bmatrix}$, $A(p_0) = \begin{bmatrix} 0 & 1 \\ 1 & 0 \end{bmatrix}$, and $B = \begin{bmatrix} 0 \\ 1 \end{bmatrix}$. Here the uncertain parameter is $p \in [-2,2]$ and it varies sinusoidally. Using (2.5) and (2.6) the upper bound of matched and mismatched components are calculated as $F_u = \begin{bmatrix} 0 & 0 \\ 0 & 0 \end{bmatrix}$ and $H = \begin{bmatrix} 4 & 4 \\ 4 & 4 \end{bmatrix}$. The parameters of (2.10) are selected as $\alpha = 1, \rho = 0.05$, and $\beta = 10$. To solve (2.29), the ARE is rewritten as

$$\hat{S}\tilde{A} + \tilde{A}^T \hat{S} + \tilde{Q} - \hat{S}\tilde{B}\tilde{R}^{-1}\tilde{B}^T \hat{S} = 0$$

with $\tilde{A} = A(p_0) = \begin{bmatrix} 0 & 1 \\ 1 & 0 \end{bmatrix}$, $\tilde{B} = \begin{bmatrix} B & \alpha(I - BB^+) \end{bmatrix} = \begin{bmatrix} 0 & 1 & 0 \\ 1 & 0 & 0 \end{bmatrix}$, $\tilde{Q} = F_u + \rho^2 H + \beta^2 I = \begin{bmatrix} 104 & 4 \\ 4 & 104 \end{bmatrix}$, and $\tilde{R} = \begin{bmatrix} I & 0 \\ 0 & \rho^2 I \end{bmatrix}$. Using the positive definite solution of the Riccati equation $\hat{S}$, the feedback control input u_2 and v are computed as

$$\begin{bmatrix} u_2 \\ v \end{bmatrix} = \begin{bmatrix} -9.9877 & -11.1232 \\ -4.9215 & -0.4994 \\ 0 & 0 \end{bmatrix} \begin{bmatrix} x_1 \\ x_2 \end{bmatrix}.$$

To compute the event-triggering conditions (2.46) and (2.57), parameters are selected as $\sigma = 0.98$, $k = 0.6$, $\lambda = (1-\sigma)k$, and $\theta = 0.1$. Here $\lambda_{min}(Q_2)$ is calculated based on (2.45). The simulation is executed for 3.5 seconds with initial states $[0.2, -0.35]^T$ and $[0.2, -0.35, 0.01]^T$ for static and dynamic cases, respectively.

Figures 2.4(b) and 2.5(b) show that the error norm is bounded by a state dependent threshold for the mismatched system. The total number of events and their corresponding positive inter-execution time for mismatched system are shown in Figs 2.4(a) and 2.5(a). For the purpose of comparison, Table 2.1 shows the inter-execution time for both event-triggered and without event-triggered control strategies. Here τ_{max}, τ_{min}, and τ_{avg} are the maximum, minimum, and average inter-event time, respectively. The scalar u_{total} represents the total number of time instants at which control input is updated during the simulation period. According to Table 2.1, the proposed event-triggered control strategy significantly reduces computation and transmission burden in the presence of parametric uncertainty. It also establishes the efficiency of dynamic event-triggering approach, as the average inter-event time is comparatively larger than the static one. The proposed event-based robust control law also ensures the existence of positive inter-event time.

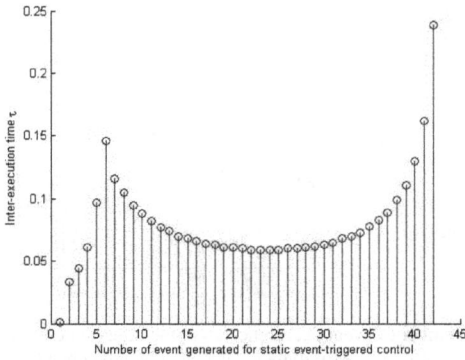

(a) Number of event occurrence and time interval between two consecutive events for static event-triggered control.

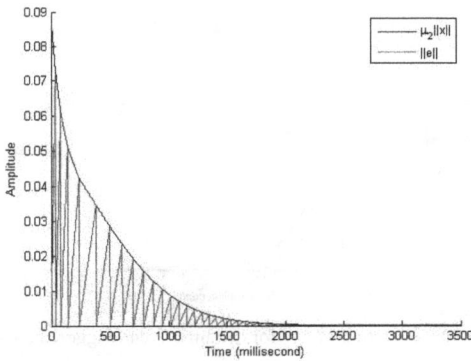

(b) Time evolution of $\|e\|$ (which is always within the threshold limit $\mu_2\|x\|$) for static event-triggered control.

Figure 2.4 Results of static event-triggered control with mismatched uncertainty.

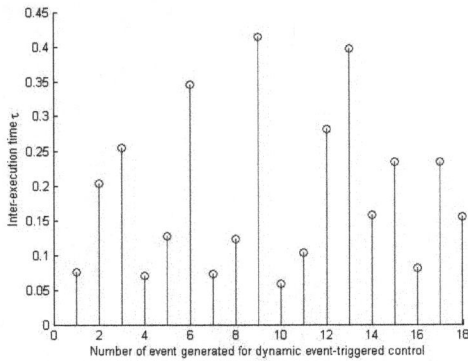

(a) Number of event occurrence and time interval between two consecutive events for dynamic event-triggered control.

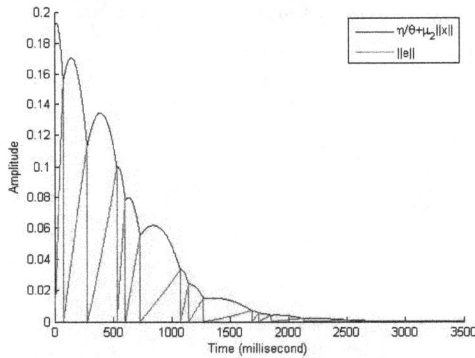

(b) Time evolution of $\|e\|$ for dynamic event-triggered control.

Figure 2.5 Results of dynamic event-triggered control with mismatched uncertainty.

2.7 SUMMARY

In this chapter, a framework of event-trigger-based robust control strategy is proposed for both matched and mismatched uncertain system. The derived control law is applicable for a wider class of linear systems in which event-triggering law is applicable. Both static and dynamic event-triggering mechanisms are adopted to design the feedback control law. For the bounded variation of parameters, the ISS of closed loop event-triggering system is proved. An analytical expression of minimum inter-event time for static and dynamic event-triggered rule are also derived. It is observed that the proposed aperiodic control law effectively reduces the total number of control input computation and information transmission to achieve a stable control law. The numerical results show the efficacy of proposed control algorithm over conventional control which does not use the event-triggered approach. The next chapter proposes an event-based robust control law for a discrete-time linear mismatched system.

3 Robust Event-triggered Control of Discrete-time Linear Systems

After studying this chapter, one should be able to: define matched and mismatched uncertain systems; design robust controller for matched and mismatched systems using LQR method; model discrete-time event-triggered control system in the presence of uncertainties; design of discrete-time robust event-triggered controller and triggering law; ensure the stability of discrete-time event-triggered systems in the face of uncertainties; validate the control approach numerically.

3.1 INTRODUCTION

In the previous chapter, a robust event-triggered control technique has been reported for a continuous-time linear system. The robust control law has been designed using the optimal control framework [80, 82]. With a periodic feedback, the existing work [80, 82] cannot be simply extended to a discrete-time system as the control input depends on the solution of discrete-time algebraic riccati equation (DT-ARE). Recently, attempt has been made to compute the robust solution for a discrete-time event-triggered system using the optimal control framework [49, 50]. But in [49, 50], it is assumed that the system is affected by matched uncertainty. This chapter avoids this assumption as there have several uncertain systems where matching condition does not hold. In this chapter, a novel discrete-time event-triggered robust control technique is proposed for a NCS, where physical system is affected by mismatched parametric uncertainty. For mathematical simplicity, we avoid other uncertainties like external disturbances, noise, etc. To derive the discrete-time robust control input, a virtual nominal system and a cost-functional are defined. The solution of the optimal control problem helps to design the stabilizing control input for the uncertain system. The ISS technique is used to derive the event-triggering condition as well as to ensure the stability of closed loop system. The chapter makes the following contributions:

(i) a robust control framework for discrete time linear system with mismatched uncertainty has been proposed in this chapter. The periodic robust control law is derived by formulating an equivalent optimal control problem. The optimal control problem is solved for a virtual nominal system with a quadratic cost-functional which depends on the upper bound of uncertainty.

(ii) The virtual nominal dynamics have two inputs u and v. The concept of virtual input v is used to derive the existence of stabilizing control input u in order to handle mismatched uncertainty. The proposed robust control law ensures asymptotic convergence of the uncertain closed loop system.

(iii) An event-triggered robust control technique is proposed for a discrete-time uncertain system, where the controller is not collocated with the system and connected through a communication network. The aim of this control law is to achieve robustness against parameter variation with event-based communication and control. The event-condition is derived from the ISS-based stability criteria.

(iv) It is shown that some of the existing results of matched system are a special case of the results presented in the work [49]-[50]. A comparison is also reported between periodic control and an event-triggered control on a numerical example to illustrate the efficacy of the proposed scheme.

Organization

The chapter is organized as follows. Section 3.2 proposes an optimal control framework to solve a robust stabilization problem for a discrete-time mismatched system with periodic feedback information. The results of event-triggered robust control law for discrete-time system is reported in Section 3.3. The numerical validation of proposed control algorithm is shown in Section 3.4. This section also compares the proposed results with the existing work reported in [49, 50]. The Section 3.5 summarizes the main contributions of this chapter.

3.2 ROBUST CONTROL DESIGN

System Description: A discrete-time uncertain linear system is described by the state equation in the form

$$x(k+1) = [A + \Delta A(p)]x(k) + [B + \Delta B(p)]u(k) \tag{3.1}$$

where $x(k) \in \mathbb{R}^n$ is the state and $u(k) \in \mathbb{R}^m$ is the control input. The matrices $A \in \mathbb{R}^{n \times n}$, $B \in \mathbb{R}^{n \times m}$ are the nominal, known constant matrices. The unknown matrices $\Delta A(p) \in \mathbb{R}^{n \times n}$ and $\Delta B(p) \in \mathbb{R}^{n \times m}$ are used to represent the system uncertainties due to bounded variation of p. The uncertain parameter vector p belongs to a predefined bounded set $\hat{P}$. Generally system uncertainties are classified as matched and mismatched uncertainties. System (3.1) suffers through the matched uncertainty if the uncertain matrices $\Delta A(p)$ and $\Delta B(p)$ satisfy the following equality

$$\Delta A(p) = B\phi_A(p) \,, \ \Delta B(p) = B\phi_B(p), \ \forall \, p \in \hat{P}. \tag{3.2}$$

In other words, $\Delta A(p)$ and $\Delta B(p)$ are in the range space of nominal input matrix B. For mismatched case, equality (3.2) does not hold. For simplification, uncertainty can be decomposed in matched and mismatched component such as

$$\Delta A(p) = S\phi_A(p) \quad = \quad BB^+ S\phi_A(p) + (I - BB^+)S\phi_A(p), \ \forall p \in \hat{P} \text{ and } S \neq 0 \quad (3.3)$$

where $BB^+ S\phi_A(p)$ is matched and $(I - BB^+)S\phi_A(p)$ is the mismatched one. Here $S \in \mathbb{R}^{n \times r}$, is a known scaling matrix and $\phi_A(p)$, $\phi_B(p)$ are unknown matrices of appropriate dimension. The matrix $B^+ (= (B^T B)^{-1} B^T)$ denotes the left pseudo inverse of matrix B [61]. The perturbations $\phi_A(p)$ and $\phi_B(p)$ are upper bounded by the positive semi-definite matrices F and H accordingly and defined as

$$\varepsilon^{-1}\phi_A^T(p)\phi_A(p) \leq F \quad (3.4)$$

$$\varepsilon^{-1}K^T \phi_B^T(p)\phi_B(p)K \leq H \quad (3.5)$$

where, scalar $\varepsilon > 0$ is a design parameter and K is a gain matrix. To stabilize (3.1), it is essential to formulate a robust control problem as stated below.

Robust control problem

Design a state feedback control law $u(k) = Kx(k)$ such that the uncertain closed loop system (3.1) is asymptotically stable for all $p \in \hat{P}$.
For simplicity, input uncertainty is not considered initially in (3.1). This reduces to following system equation

$$x(k+1) = (A + \Delta A(p))x(k) + Bu(k) \quad (3.6)$$

with the nominal dynamics

$$x(k+1) = Ax(k) + Bu(k). \quad (3.7)$$

In order to stabilize (3.6), the robust controller gain K is designed through an optimal control approach.

Optimal control approach

The essential idea is to design an optimal control law for a virtual nominal dynamics which minimizes a modified cost-functional, J. The cost-functional is called modified as it depends on the upper-bound of uncertainty. A virtual system (3.8) is defined by adding an extra term $\alpha(I - BB^+)Sv(k)$ in equation (3.7). The virtual dynamics and cost-functional for (3.6) are given below:

$$x(k+1) = Ax(k) + Bu(k) + \alpha(I - BB^+)Sv(k) \quad (3.8)$$

$$J(k) = \frac{1}{2}\sum_{k=0}^{\infty} \left\{ x(k)^T (\beta^2 I + Q + F)x(k) + u(k)^T R_1 u(k) + v(k)^T R_2 v(k) \right\} \quad (3.9)$$

where $Q \geq 0$, $R_1 > 0$, $R_2 > 0$ and scalers α, β are design parameters. The derived optimal input for virtual system is proved to be a robust input for the original uncertain system.

> **Remark 11** *Here $u = Kx$ is the stabilizing control input and $v(k) = Lx(k)$ is a virtual input. The input v is called virtual since it is not used directly to stabilize (3.6). But it helps indirectly to design K for mismatched system. The usefulness of L in the context of event-triggered control is discussed in Section 3.3.*

The robust control law for (3.6) is designed by minimizing (3.9) for the virtual system (3.8). The results are stated in the form a theorem.

> **Theorem 6** *Suppose there exist a scalar $\varepsilon > 0$ and positive definite solution $P > 0$ of equation (3.10)*
>
> $$A^T \{P^{-1} + BR_1^{-1}B^T + \alpha^2(I - BB^+)SR_2^{-1}S^T(I - BB^+)^T\}^{-1}$$
> $$A - P + Q + F + \beta^2 I = 0 \qquad (3.10)$$
>
> *with*
>
> $$(\varepsilon^{-1}I - S^T PS) > 0, \qquad (3.11)$$
> $$\varepsilon^{-1}\phi_A^T(p)\phi_A(p) \leq F. \qquad (3.12)$$
>
> *Moreover, the controller gains K and L are computed as*
>
> $$K = -R_1^{-1}B^T\{P^{-1} + BR_1^{-1}B^T + \alpha^2(I - BB^+)SR_2^{-1}S^T(I - BB^+)^T\}^{-1}A, \qquad (3.13)$$
>
> $$L = -\alpha R_2^{-1}S^T(I - BB^+)^T$$
> $$\{P^{-1} + B^T R_1^{-1}B + \alpha^2(I - BB^+)SR_2^{-1}S^T(I - BB^+)^T\}^{-1}A. \qquad (3.14)$$
>
> *The matrix K is the robust controller gain for (3.6), if it satisfies the following matrix inequality*
>
> $$Q_1 = (Q + \beta^2 I + K^T R_1 K + L^T R_2 L + M^T P^{-1}M) - A_c^T(P^{-1} - \varepsilon SS^T)^{-1}A_c \geq 0 \qquad (3.15)$$
>
> *where $A_c = A + BK$ and matrix*
>
> $$M = \{P^{-1} + BR_1^{-1}B^T + \alpha^2(I - BB^+)SR_2^{-1}S^T(I - BB^+)^T\}^{-1}A. \qquad (3.16)$$

To prove the above theorem some intermediate results are stated in the form of two lemmas.

> **Lemma 4** *(Garcia et al. [51]) Let P be a positive definite solution of (3.10). Then there exists a scalar $\varepsilon > 0$ such that*
>
> $$A_c^T PS\phi_A + \phi_A^T S^T PA_c + \phi_A^T S^T PS\phi_A \leq A_c^T PS(\varepsilon^{-1}I - S^T PS)^{-1}S^T PA_c + \varepsilon^{-1}\phi_A^T \phi_A \qquad (3.17)$$

with

$$(\varepsilon^{-1}I - S^T PS) > 0 \tag{3.18}$$

Proof *Let* $W = (\varepsilon^{-1}I - S^T PS)^{-\frac{1}{2}} S^T PA_c - (\varepsilon^{-1}I - S^T PS)^{\frac{1}{2}} \phi_A$ *where* $(\varepsilon^{-1}I - S^T PS) > 0$, *then* $W^T W$ *is always positive for any selection of* W. *After multiplication, the simplified form of* $W^T W > 0$ *is*

$$A_c^T PS(\varepsilon^{-1}I - S^T PS)^{-1} S^T PA_c - A_c^T PS\phi_A - \phi_A^T S^T PA_c + \phi_A^T (\varepsilon^{-1}I - S^T PS)\phi_A \geq 0. \tag{3.19}$$

After Rearranging (3.19), following inequality is obtained

$$A_c^T PS\phi_A + \phi_A^T S^T PA_c + \phi_A^T S^T PS\phi_A \leq A_c^T PS(\varepsilon^{-1}I - S^T PS)^{-1} S^T PA_c + \varepsilon^{-1} \phi_A^T \phi_A. \tag{3.20}$$

The inequality (3.20) holds $\forall \ (\varepsilon^{-1}I - S^T PS) > 0$ *and this ends the proof.*

Lemma 5 *Let* $P > 0$ *be a solution of (3.10) which satisfies the equation (3.44). Using controller gain (3.13) and (3.14), the following inequality holds*

$$A^T \{P^{-1} + BR_1^{-1}B^T + \alpha^2(I - BB^+)SR_2^{-1}S^T(I - BB^+)^T\}^{-1}A$$
$$= (K^T R_1 K + L^T R_2 L + M^T P^{-1}M). \tag{3.21}$$

Proof *Let* $M = \{P^{-1} + BR_1^{-1}B^T + \alpha^2(I - BB^+)SR_2^{-1}S^T(I - BB^+)^T\}^{-1}A$, *then the following equation is simplified as*

$$A^T \{P^{-1} + BR_1^{-1}B^T + \alpha^2(I - BB^+)SR_2^{-1}S^T(I - BB^+)^T\}^{-1}A = A^T \{P^{-1} + BR_1^{-1}B^T$$
$$+ \alpha^2(I - BB^+)SR_2^{-1}S^T(I - BB^+)^T\}^{-1}P^{-1}\{P^{-1} + BR_1^{-1}B^T + \alpha^2(I - BB^+)SR_2^{-1}S^T(I$$
$$- BB^+)^T\}^{-1}A + A^T \{P^{-1} + BR_1^{-1}B^T + \alpha^2(I - BB^+)SR_2^{-1}S^T(I - BB^+)^T\}^{-1}(BR_1^{-1}B^T)$$
$$\{P^{-1} + BR_1^{-1}B^T + \alpha^2(I - BB^+)SR_2^{-1}S^T(I - BB^+)^T\}^{-1}A + A^T \{P^{-1} + BR_1^{-1}B^T + \alpha^2(I$$
$$- BB^+)SR_2^{-1}S^T(I - BB^+)^T\}^{-1}\alpha^2(I - BB^+)SR_2^{-1}S^T(I - BB^+)^T \{P^{-1} + BR_1^{-1}B^T$$
$$+ \alpha^2(I - BB^+)SR_2^{-1}S^T(I - BB^+)^T\}^{-1}A. \tag{3.22}$$

Substituting the value of gain matrices K *and* L *from (3.13) and (3.14), the above equation can be written as*

$$A^T \{P^{-1} + BR_1^{-1}B^T + \alpha^2(I - BB^+)SR_2^{-1}S^T(I - BB^+)^T\}^{-1}A$$
$$= M^T P^{-1}M + K^T R_1 K + L^T R_2 L. \tag{3.23}$$

Proof of Theorem 6 *The proof is divided in two parts. At first, we solve an optimal control problem to minimize (3.9) for the nominal system (3.8). For this purpose* *the optimal input u and v should minimize the Hamiltonian H, that means* $\dfrac{\partial H}{\partial u} = 0$

and $\dfrac{\partial H}{\partial v} = 0$. *After applying discrete-time LQR methods, the Riccati equation (3.10) and controller gains (3.13), (3.14) are achieved [91]. To prove the stability of uncertain system, let* $V(x) = x(k)^T Px(k)$ *be a Lyapunov function for (3.6). Then applying (3.41), the time difference of* $V(x)$ *is*

$$\Delta V = x(k)^T [A_c^T PA_c + A_c^T PS\phi_A + \phi_A^T S^T PA_c + \phi_A^T S^T PS\phi_A - P]x(k) \tag{3.24}$$

where $A_c = (A + BK)$. *Using matrix inversion lemma, following is achieved [61]*

$$(P^{-1} - \varepsilon SS^T)^{-1} = P + PS(\varepsilon^{-1}I - S^T PS)^{-1}S^T P. \tag{3.25}$$

Using (3.10), (3.17), and (3.25) in (3.24), ΔV *is simplified as*

$$\begin{aligned}\Delta V \leq &x(k)^T [A_c^T (P^{-1} - \varepsilon SS^T)^{-1}A_c - A^T \{P^{-1} + BR_1^{-1}B^T \\ &+ \alpha^2 (I - BB^+)SR_2^{-1}S^T (I - BB^+)^T\}^{-1}A]x(k) \\ &- x(k)^T [F - \varepsilon^{-1}\phi_A^T \phi_A + Q + \beta^2 I]x(k).\end{aligned} \tag{3.26}$$

Now applying Lemma 5, (3.26) is written as

$$\Delta V \leq -x(k)^T Q_1 x(k). \tag{3.27}$$

The inequality (3.27) will be negative semi-definite if and only if equation (3.15) is satisfied.

In order to tackle uncertainty in input matrix, a cost-functional is defined as

$$J(k) = \frac{1}{2}\sum_{k=0}^{\infty}\left\{x(k)^T (\beta^2 I + Q + 2F + 2H)x(k) + u(k)^T R_1 u(k) + v(k)^T R_2 v(k)\right\}. \tag{3.28}$$

The robust control law for (3.1) can be computed by solving an optimal control problem using (3.28) and (3.8). The results are stated as follows.

Theorem 7 *Suppose there exist a scalar* $\varepsilon > 0$ *and a positive definite solution P of Riccati equation*

$$\begin{aligned}A^T \{P^{-1} + BR_1^{-1}B^T + \alpha^2 (I - BB^+)SR_2^{-1}S^T (I \\ - BB^+)^T\}^{-1}A - P + Q + 2F + 2H + \beta^2 I = 0\end{aligned} \tag{3.29}$$

with

$$(\varepsilon^{-1}I - S^T PS) > 0, \tag{3.30}$$
$$\varepsilon^{-1}\phi_A^T \phi_A \leq F, \tag{3.31}$$
$$\varepsilon^{-1}K^T \phi_B^T \phi_B K \leq H. \tag{3.32}$$

> *Using the solution P, the controller gain (3.13) [a] is the robust solution of (3.1) if it satisfy the following inequality:*
>
> $$Q_1 = (Q + \beta^2 I + K^T R_1 K + L^T R_2 L + M^T P^{-1} M)$$
> $$- A_c^T (P^{-1} - \varepsilon S S^T)^{-1} A_c \geq 0. \qquad (3.33)$$
>
> ---
>
> [a]The gain matrices K and L for Theorem 7 are structurally similar to the matrices used in Theorem 6. However, these two matrices are different as they depend on the solution of different Riccati equation.

Proof *If $V(x(k)) = x(k)^T P x(k)$ is a Lyapunov function for (3.1) then ΔV along the direction of x is*

$$\Delta V = x(k+1)^T P x(k+1) - x(k)^T P x(k). \qquad (3.34)$$

Using (3.1), the ΔV is expressed as

$$\Delta V = x(k)^T [\{(A+BK)^T P + (\phi_A + \phi_B K)^T S^T P\}\{(A+BK) + S(\phi_A + \phi_B K)\} - P]x(k).$$

Denoting $A_c = (A+BK)$ and $\Delta = (\phi_A + \phi_B K)$, the above equality reduces to

$$\Delta V = x(k)^T \{A_c^T P A_c + A_c^T P S \Delta + \Delta^T S^T P A_c + \Delta^T S^T P S \Delta - P\}x(k). \qquad (3.35)$$

Using similar concept of Lemma 4, (3.35) can be written as

$$\Delta V \leq x(k)^T \{A_c^T P A_c + A_c^T P S(\varepsilon^{-1} I - S^T P S)^{-1} S^T P A_c + \varepsilon^{-1} \Delta^T \Delta - P\}x(k) \qquad (3.36)$$

where $\varepsilon > 0$ and satisfy $(\varepsilon^{-1} I - S^T P S)^{-1} > 0$.
Fact: *For vectors a and b with $\lambda_1 > 0$, the following equality always holds:*

$$\lambda_1^{-1}(a+b)^T(a+b) \leq \frac{2}{\lambda_1} a^T a + \frac{2}{\lambda_1} b^T b. \qquad (3.37)$$

Using (3.37), the upper bound of $\varepsilon^{-1} \Delta^T \Delta$ is defined as

$$\varepsilon^{-1} \Delta^T \Delta \leq \frac{2}{\varepsilon} \phi_A A^T \phi_A + \frac{2}{\varepsilon} K^T \phi_B^T \phi_B K. \qquad (3.38)$$

After substituting (3.29) and (3.38) and applying Lemma 4, equation (3.36) is simplified as

$$\Delta V \leq x(k)^T [A_c^T P A_c + A_c^T P S(\varepsilon^{-1} I - S^T P S)^{-1} S^T P A_c$$
$$- (F - \frac{2}{\varepsilon} \phi_A^T \phi_A) - (H - \frac{2}{\varepsilon} K^T \phi_B^T \phi_B K)$$
$$- Q - \beta^2 I - A^T \{P^{-1} + B R_1^{-1} B^T + \alpha^2 (I - BB^+) S R_2^{-1} S^T (I - BB^+)^T\} A]x(k).$$
$$(3.39)$$

Using (3.31) and (3.32) the equality (3.39) reduces to following inequality:

$$\Delta V \leq x(k)^T [A_c^T P A_c + A_c^T P(\varepsilon^{-1} I - S^T P S)^{-1} P A_c - Q - \beta^2 I - A^T \{P^{-1} + B R_1^{-1} B^T$$
$$+ \alpha^2 (I - BB^+) S R_2^{-1} S^T (I - BB^+)^T\} A]x(k).$$

Using Lemma 5, the ΔV is expressed as

$$\Delta V \leq -x(k)^T \{(Q + \beta^2 I + K^T R_1 K + L^T R_2 L + M^T P^{-1} M) - A_c^T (P^{-1} - \varepsilon S S^T)^{-1} A_c\} x(k).$$

The ΔV will be negative definite if condition (3.33) is satisfied.

The results of Theorem 6 and 7 do not consider any communication constraint in realizing the control law. So we formulate a robust control problem for an uncertain system with event-triggered control input. The block diagram of proposed robust control technique is shown in Fig. 3.1. It has three primary parts, namely, system

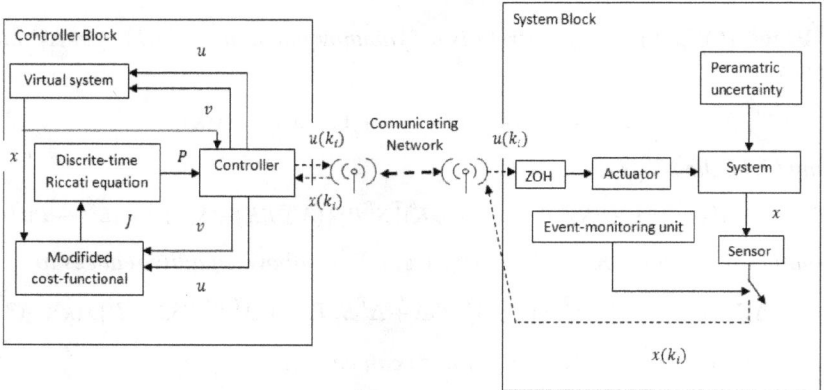

Figure 3.1 Block diagram of the proposed robust control law with limited communication. The aperiodic transmission of information is represented by dotted lines.

block, controller block and a bandwidth limited communicating network between system and controller. The states of system are periodically measured by the sensor which is collocated with the system. The sensor is connected with the controller via a communication network. An event-monitoring unit verifies a state-dependent event condition periodically and transmit state information to the controller only when the event-condition is satisfied. The robust controller gain K with eventual state information, $x(k_i)$ which is received from uncertain system is used to generate the event-triggered control law $u(k_i) = Kx(k_i)$ to stabilize (3.1) or (3.6). Here k_i denotes the latest event-triggering instant and the control input is updated at an aperiodic discrete-time instant $k_0, k_1, k_2 \cdots$. A zero-order-hold (ZOH) is used to hold the last transmitted control input until the next input is transmitted. Here the actuator is collocated with the system and the actuating control law is assumed to change instantly with the transmission of control input.

The discrete-time linear uncertain system (3.1) with event-triggered input

$$u(k_i) = Kx(k_i), \quad k \in [k_i, k_{i+1}) \tag{3.40}$$

are written as

$$x(k+1) = (A + \Delta A(p))x(k) + Bu(k_i). \tag{3.41}$$

From [40], (3.41) can be modeled as

$$x(k+1) = (A + \Delta A + BK)x(k) + BKe(k). \tag{3.42}$$

The variable $e(k)$ is named as measurement error. It is used to represent the eventual state information $x(k_i)$ in the form

$$e(k) = x(k_i) - x(k) \quad k \in [k_i, \ k_{i+1}). \tag{3.43}$$

Problem Statement: Design a feedback control law to stabilize the uncertain discrete-time event-triggered system (3.41) such that the closed loop system is ISS with respect to its measurement error $e(k)$.

Proposed solution: This problem is solved in two different steps. Firstly, the controller is designed by adopting the results of optimal control theory and secondly an event-triggering rule is derived to make (3.41) ISS. The design procedure of controller gains, based on Theorems 6 and 7, are already discussed in Section 3.2. The event-triggering law is derived from the Definition 9 and 10 assuming an ISS Lyapunov function $V(x) = x^T P x$. The design procedure of event-triggering condition is discussed in Section 3.3.

3.3 MAIN RESULTS

This section discusses the design procedure of event-triggering condition and stability proof of (3.41) in the presence of uncertainty. The results are stated in the form of a theorem.

Theorem 8 *Let $P > 0$ be a solution of the Riccati equation (3.10) for a scalar $\varepsilon > 0$ and satisfy the following inequalities*

$$(\varepsilon^{-1}I - S^T PS) \quad > 0 \tag{3.44}$$

$$\varepsilon^{-1}\phi_A^T(p)\phi_A(p) \quad \leq F/2 \tag{3.45}$$

$$Q_1 = Q + \beta^2 I + K^T R_1 K + L^T R_2 L + M^T P^{-1} M - A_c^T (P^{-1} - \varepsilon S S^T)^{-1} A_c \geq 0 \tag{3.46}$$

where controller gains K and L are computed by (3.13), (3.14). The event-triggered control law (3.40) ensures the ISS of (3.41) $\forall p \in \hat{P}$ if the input is updated based on following triggering-sequence

$$k_0 = 0, k_{i+1} = \inf \left\{ \ k \in \mathbb{I} \mid k \geq k_i \wedge (\mu_1 \|x(k)\|^2 - \|e(k)\|^2 \leq 0) \ \right\}. \tag{3.47}$$

Moreover the design parameter μ_1 is explicitly defined as

$$\mu_1 = \frac{\sigma \lambda_{min}^2(Q_1)}{2\lambda_{min}(Q_1)\|K^T B^T (P^{-1} - \varepsilon S S^T)^{-1} BK\| + 4\|A_c^T PBK\|^2} \tag{3.48}$$

where scalar $\sigma \in (0,1)$.

Proof of Theorem 8 *Assume $V(x)$ as a ISS Lyapunov function for (3.41) and $A_c = (A + BK)$. Using Lemma 4 and the solution P of Riccati equation (3.10), the time difference of $V(x)$ $[\Delta V = V(x(k+1)) - V(x(k))]$ along the direction of state trajectory of (3.41) is simplified as*

$$
\begin{aligned}
\Delta V = &-x^T(Q+\beta^2 I+F-2\varepsilon^{-1}\phi_A^T\phi_A)x+2x^TA_c^TPBKe\\
&+e^TK^TB^T(P^{-1}-\varepsilon SS^T)^{-1}BKe\\
&+x^T[A_c^T(P^{-1}-\varepsilon SS^T)^{-1}A_c-A^T\{P^{-1}+BR_1^{-1}B^T\\
&+\alpha^2(I-BB^+)SR_2^{-1}S^T(I-BB^+)^T\}^{-1}A]x.
\end{aligned}
\tag{3.49}
$$

Using Lemma 5 and (3.45), equation (3.49) is simplified as

$$
\Delta V \leq -x^T Q_1 x + 2x^T A_c^T PBKe + e^T K^T B^T (P^{-1} - \varepsilon SS^T)^{-1} BKe. \tag{3.50}
$$

The upper bound of (3.50) is derived as

$$
\Delta V \leq -\frac{\lambda_{min}(Q_1)}{2}\|x\|^2 + \left(\frac{2\|A_c^T PBK\|^2}{\lambda_{min}(Q_1)} + \left\|K^T B^T(P^{-1} - \varepsilon SS^T)^{-1}BK\right\|\right)\|e\|^2. \tag{3.51}
$$

Using Definition 10, (3.51) ensures ISS of (3.41) if conditions (3.44)–(3.46) hold. Applying the ISS stability property from Definition 10 on (3.51), the event-triggering condition (3.47) is achieved.

An algorithmic representation of discrete-time event-triggered control for mismatched system is introduced below: The proposed robust control framework considers the general system uncertainty, which includes both matched and mismatched component. Without the mismatched part, the system (3.6) reduces to a matched system (defined in (3.2)), i.e. $\Delta A = B\phi_A(p)$. Moreover, due to the absence of the mismatched part, the virtual control input v is not necessary in (3.8). As a special case of Theorem 8, the Corollary 1 is introduced for matched systems.

Corollary 1 *Suppose there exist a scalar $\varepsilon > 0$ and a positive definite solution P of Riccati equation*

$$
A^T\{P^{-1} + BR_1^{-1}B^T)\}^{-1}A - P + Q + F + \beta^2 I = 0 \tag{3.52}
$$

with following conditions

$$
\varepsilon^{-1}\phi_A^T\phi_A \leq F/2 \tag{3.53}
$$

$$
(\varepsilon^{-1}I - B^T PB) > 0 \tag{3.54}
$$

$$
Q_2 = Q + \beta^2 I + K^T R_1 K + M_1^T P^{-1} M_1 - A_c^T(P^{-1} - \varepsilon BB^T)^{-1}A_c \geq 0 \tag{3.55}
$$

where controller gain $K = -R_1^{-1}B^T\{P^{-1} + BR_1^{-1}B^T\}^{-1}A$ and $M_1 = \{P^{-1} + BR_1^{-1}B^T\}^{-1}A$. The controller gain K ensures the ISS of matched system if the *control input is actuated based on the following event-triggering sequence*

$$k_0 = 0, k_{i+1} = inf\left\{ \ k \in \mathbb{I} \mid k \geq k_i \wedge (\mu_2 \|x(k)\|^2 - \|e(k)\|^2 \leq 0) \ \right\}. \quad (3.56)$$

The parameter μ_2 is derived as

$$\mu_2 = \frac{\frac{1}{2}\sigma\lambda_{min}(Q_2)}{\|K^T B^T (P^{-1} - \varepsilon BB^T)^{-1} BK\| + \frac{2}{\lambda_{min}(Q)} \|A_c^T PBK\|^2} \quad (3.57)$$

where scalar $\sigma \in (0,1)$.

Algorithm 2 Robust Event-triggered control

1: Initialization: $k \Leftarrow 0$, $x \Leftarrow x_0$, $x(k_i) \Leftarrow x_0$.
2: Select any value of $\varepsilon > 0$.
3: For a given A, B, F, σ, compute P, K, L [from (3.10), (3.13), and (3.14)].
4: Check the conditions (3.44), (3.46), and (3.45).
5: **if** Conditions (3.44), (3.46), and (3.45) hold **then**
6:     Go to line 11
7: **else**
8:     Go to line 2
9: **end if**
10: Compute $\|x(k)\|$, $\|e(k)\|$ and μ_1 using (3.48).
11: **if** $\|e\|^2 \geq \mu_1 \|x\|^2$ **then**
12:     Transmit $x(k_i)$ from system to controller.
13:     Compute and update $u(k_i) = Kx(k_i)$ in (3.40).
14: **else**
15:     $u(k) = u(k_{i-1})$
16: **end if**
17: Return to line 10

Proof *Considering the similar Lyapunov function for (3.6), the ΔV is derived as*

$$
\begin{aligned}
\Delta V &= [(A + B\phi_A + BK)x + BKe]^T P[(A + B\phi_A + BK)x + BKe] - x^T Px \\
&= [x^T (A_c + B\phi_A)^T P(A_c + B\phi_A)x + e^T K^T B^T P(A_c + B\phi_A)x + x^T (A + B\phi_A)^T PBKe] \\
&\quad + e^T K^T B^T PBKe - x^T Px.
\end{aligned} \quad (3.58)
$$

Here notation A_c is used to denote $(A + BK)$. Using Lemma 4, the equality (3.58) is simplified as

$$
\begin{aligned}
\Delta V &\leq x^T (A_c^T (P^{-1} - \varepsilon BB^T)^{-1} A_c + 2\varepsilon^{-1} \phi_A^T \phi_A - P)x + \{K^T B^T (P^{-1} - \varepsilon BB^T)^{-1} BK + \\
&\quad \frac{2}{\lambda_{min}(Q_2)} \|A_c^T PBK\|^2\} \|e\|^2 .
\end{aligned} \quad (3.59)
$$

Now to prove the ISS of (3.6), applying Lemma 5, the above inequality reduces to

$$\Delta V \leq -x^T(Q + \beta^2 I + K^T R_1 K + M_1^T P^{-1} M_1 - A_c^T (P^{-1} - \varepsilon BB^T)^{-1} A_c)x + \{K^T B^T (P^{-1}$$
$$-\varepsilon BB^T)^{-1} BK + \frac{2}{\lambda_{min}(Q_2)} \left\| A_c^T PBK \right\|^2 \} \|e\|^2. \tag{3.60}$$

where $M_1 = (P^{-1} + BR_1^{-1} B^T)^{-1} A$. The inequality (3.60) is written in the form of (1.43) to ensure ISS for (3.6) if inequality (3.55) is satisfied. Using (3.60) the event-triggering condition (3.56) for (3.41) is achieved.

3.3.1   UNCERTAINTY IN INPUT MATRIX

With event-triggered control input (3.40), the uncertain system (3.6) turns as

$$x(k+1) = \left[A + \Delta A(p)\right] x(k) + \left[B + \Delta B(p)\right] u(k_i). \tag{3.61}$$

From [40], (3.61) reduces as

$$x(k+1) = (A + BK + \Delta A + \Delta BK)x(k) + (BK + \Delta BK)e(k) \tag{3.62}$$

where the variable $e(k)$ is defined in (3.43). With both state and input uncertainties, the stability criteria and event-triggering condition for (3.61) are stated below.

Theorem 9 *Let $P > 0$ be a solution of the Riccati equation (3.29) for a scalar $\varepsilon > 0$ and satisfy the following inequalities*

$$(\varepsilon^{-1} I - S^T PS) > 0, \tag{3.63}$$

$$\varepsilon^{-1} \phi_A^T \phi_A \leq F/2, \tag{3.64}$$

$$\varepsilon^{-1} K^T \phi_B^T \phi_B K \leq H/2, \tag{3.65}$$

$$Q_1 = Q + \beta^2 I + K^T R_1 K + L^T R_2 L + M^T P^{-1} M - A_c^T (P^{-1} - \varepsilon SS^T)^{-1} A_c > 0 \tag{3.66}$$

where controller gains K and L are computed by (3.13) and (3.14). The event-triggered control law (3.40) ensures ISS of (3.61) $\forall p \in \hat{P}$ if the input is updated through the following event-triggering sequence

$$k_0 = 0, k_{i+1} = \inf \left\{ k \in \mathbb{I} \mid k \geq k_i \wedge (\mu_3 \|x(k)\|^2 - \|e(k)\|^2 \leq 0) \right\}. \tag{3.67}$$

The design parameter μ_3 is defined as

$$\mu_3 = \frac{\frac{\sigma \lambda_{min}(Q_1)}{2}}{\left\| \hat{\Delta}^T \left[(P^{-1} - \varepsilon SS^T)^{-1} + \frac{2}{\lambda_{min}(Q_1)} PA_c A_c^T P \right] \hat{\Delta} \right\|}$$

where parameters $\sigma \in (0,1)$ and $\hat{\Delta} = (BK + S\Delta BK)$.

Proof *The time difference of $V(x)$ along the direction of (3.61) is*

$$\Delta V = x(k+1)^T P x(k+1) - x(k)^T P x(k). \tag{3.68}$$

After substituting $x(k+1)$ from (3.61), the ΔV is reduces to

$$\Delta V = \{x(k)^T (A_c^T P + (\phi_A + \phi_B K)^T S^T P + e^T (BK + S\phi_B K) P\} \{(A_c + S(\phi_A + \phi_B K)) x(k) \\ + (BK + S\phi_B K) e(k)\}.$$

Now $(A + BK)$, $(\phi_A + \phi_B K)$, and $(BK + S\phi_b K)$ are denoted by A_c, Δ, and $\tilde{\Delta}$, which simplify the above equality into the following form

$$\Delta V = x(k)^T \left[Ac^T PA_c + A_c^T PS\Delta + \Delta^T S^T PA_c + \Delta^T S^T PS\Delta \right] x(k) + (x^T \Delta^T S^T P \tilde{\Delta} e \\ + e^T \tilde{\Delta}^T PS\Delta x + e^T \tilde{\Delta}^T P\tilde{\Delta} e) + 2x^T A_c^T P \tilde{\Delta} e - x(k)^T Px(k). \tag{3.69}$$

Using Lemma 4 the above equality is written as

$$\Delta V \leq x(k)^T A_c^T (P - \varepsilon SS^T)^{-1} A_c x(k) + 2\varepsilon^{-1} x(k)^T \Delta^T \Delta x(k) + e^T \tilde{\Delta}^T (P^{-1} - \varepsilon SS^T)^{-1} \tilde{\Delta} e \\ + 2x(k)^T A_c^T P \tilde{\Delta} e - x^T (k) Px(k). \tag{3.70}$$

After further simplification and applying Lemma 5 in (3.70), the following is obtained

$$\Delta V \leq -\frac{\lambda_{min}(Q_1)}{2} \|x\|^2 + e^T \tilde{\Delta}^T \left[(P^{-1} - \varepsilon SS^T)^{-1} + \frac{2}{\lambda_{min}(Q_1)} PA_c A_c^T P \right] \tilde{\Delta} e \tag{3.71}$$

where $Q_1 = Q + \beta^2 I + K^T R_1^{-1} K + L^T R_2 L + M^T P^{-1} M - A_c^T (P - \varepsilon SS^T)^{-1} A_c$. Using Definition 10, the equation (3.71) ensures ISS of (3.61) if conditions (3.63)–(3.66) hold. Applying the ISS stability property from Definition 10 on (3.71), the event-triggering condition (3.67) for (3.61) is achieved.

Remark 12 *The matrices F and H are used to define the upper bound of uncertain matrices $\phi_A^T \phi_A$ and $\phi_B^T \phi_B$. After fixing the controller gains K and L, it is observed from (3.12) and (3.45) [or (3.31)–(3.32) and (3.64)–(3.65)] that the ISS property can be ensured for the event-triggered system, if the upper bound of uncertainty is $\frac{1}{\sqrt{2}}$ times of the periodic one.*

Remark 13 *The event-triggering law (3.47) [or (3.67)] for (3.41) [or for (3.61)] is designed. The designed parameter μ_1 [or μ_2] depends on the virtual gain L. Therefore L has direct influence to design the robust stabilizing controller gain K as well as on event-triggering law. The selection of design parameter β and ε in (3.9) [or (3.28)] has a greater significance on system stability due to the fact that the positive definiteness of (3.15) depends on these parameters.*

Remark 14 *The proposed event-triggering condition [mentioned in (3.47)] for (3.41) [or (3.67) for (3.61)] is monitored periodically for a fixed sampling period. In the event-triggered control of a discrete-time system, the sampling period directly provides the lower bound of inter-event time [57]. A large lower bound of inter-event time causes less data transmission, which results in poor closed loop performance. Hence the selection of sampling period for an event-triggered control of a discrete-time system plays a crucial role as it decides the usage of communication resources by maintaining the system stability and performance. The joint design of event-triggering rule and the controller gain is a difficult problem. In this chapter, for a choice of the sampling period, the designed controller gain (3.13) ensures the stability of the closed loop uncertain system (3.6) [or (3.1)] [as mentioned in Theorem 6 and 7]]. Then the proposed event-triggering rule (3.47) for (3.41) [or (3.67) for (3.61)] is designed for a specific value of the design parameter $\sigma \in (0, 1)$. The value of σ regulates the system performance and the usage of communication resources.*

3.3.2 COMPARISON WITH EXISTING RESULTS

This subsection compares the main contribution of this chapter with the results reported in [49, 50]. To compare with [49], the mismatched part of uncertainty is neglected. The Riccati equation mentioned in [49] is similar to proposed Riccati equation (3.10) in the presence of matched uncertainty. The only difference is the presence of additional term $\beta^2 x^T x$ in the cost-functional (3.10). Without this extra term the Riccati equation (3.10) and controller gain (3.13) reduce to a similar form as mentioned in [49]. So the results of [49] can be recovered as a special case of proposed results.

Moreover in Theorem 1 of [49, 50], an event-triggering condition is stated for matched system which depends on uncertain matrix $\phi_A(p)$. Implementation of this event-triggering condition is not realistic as matrix $\phi_A(p)$ is unknown. However in this chapter, the proposed event triggering condition discussed in Theorem 8 is independent of uncertainty $\phi_A(p)$. It directly depends on the controller gain K. The triggering condition for matched system is also independent of uncertain matrix explicitly as seen from (3.57).

3.4 NUMERICAL EXAMPLES AND COMPARATIVE STUDIES

To demonstrate the effectiveness of the proposed algorithm, a rotating base inverted pendulum is considered from [102, 46]. For stabilization of this system in the presence of uncertainty, the closed loop feedback control law is realized over the wireless network. The control law is designed excluding the effect of network properties. The discrete-time linearized model of rotating base pendulum is computed from a continuous model for a sampling period

$T = 0.005s$, where system matrices $A = \begin{bmatrix} 1.0008 & 0.005 & 0 & 0 \\ 0.3164 & 1.008 & 0 & 0 \\ -0.0004 & 0 & 1 & 0.005 \\ -0.1666 & -0.0004 & 0 & 1 \end{bmatrix}$, $B =$

$\begin{bmatrix} -0.0065 & -2.6043 & 0.0101 & 4.0210 \end{bmatrix}^T$. The matrices A and B depend on the selection of sampling period. The discretization method outlined in [23, 125] is adopted for the computation of A and B matrices. Here the system is affected by mismatched uncertainty and therefore the results of [49]–[50] are not applicable. To solve (3.10), different matrices are $F = \begin{bmatrix} 48.40 & 24.20 & 9.68 & 7.26 \\ 24.20 & 12.10 & 4.84 & 3.63 \\ 9.68 & 4.84 & 1.93 & 1.45 \\ 7.26 & 3.63 & 1.45 & 1.08 \end{bmatrix}$ and Q, R_1, R_2 are identity matrices with appropriate dimension. Design parameters $\varepsilon = 0.001$, $\alpha = 0.01$, $\beta = 2$, and $\sigma = 0.8$ are used. To design event-triggering condition the numerical value of μ_1 is found to be 0.018 (using (3.48)). The system uncertainty (ΔA) is defined by two matrices $S = \begin{bmatrix} 0.0065 & -2.5638 & 0.0182 & 3.9805 \end{bmatrix}^T$ and $\phi(p) = p \times sin(0.6k) \begin{bmatrix} 0.22 & 0.11 & 0.04 & 0.03 \end{bmatrix}$ where $p \in \pm 1.1$. For the purpose of simulation the value of p is selected as 0.7, which will satisfy the condition $\varepsilon^{-1} \phi_A^T \phi_A \leq F/2$. The simulation is carried-out in Matlab for a total time steps of $k = 1200$ with a sampling interval $T = 0.005s$. The initial value of states are $x = \begin{bmatrix} 0.1 & -0.5 & 0.2 & -0.5 \end{bmatrix}^T$. The controller gains K and L are computed using (3.13), (3.14) and their numerical values are $K = \begin{bmatrix} 4.98 & 0.826 & 0.215 & 0.265 \end{bmatrix}$ and $L = \begin{bmatrix} -1.494 & -0.231 & -0.212 & -0.143 \end{bmatrix}$.

Table 3.1

Comparative Results of Event-triggered and Periodic Robust Control Approach.

Control mechanism	$\tau_{max}(sec.)$	$\tau_{min}(sec.)$	u_{total}
Periodic control	0.005	0.005	1200
Event-triggered control (For $S \neq B$)	0.03	0.005	475
Event-triggered control (For $S = B$)	0.03	0.005	467

ANALYSIS OF SIMULATION RESULTS

Figure 3.2 shows the convergence of states for event-triggered and periodic input. A zoomed view of Fig. 3.2 is also shown in the same figure and it is observed that the performance of event-triggered controller is poorer as compared to periodic one due to limited availability of stabilizing input. By adjusting the numerical value of σ one can achieve the acceptable performance. For the purpose of comparison, Table 3.1 shows the number of transmission requirements in total run-time and inter-execution time for both event-triggered and periodic control strategy. Here τ_{max} and τ_{min} are the maximum and minimum number of inter-event intervals, respectively. The scalar

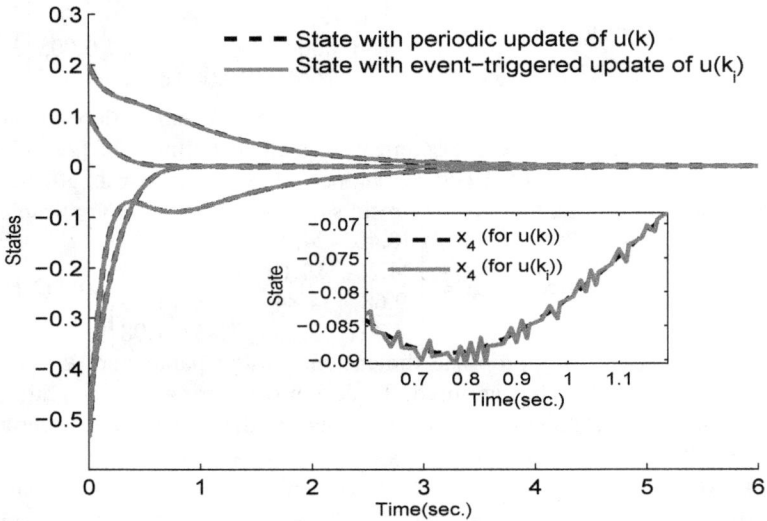

Figure 3.2 Convergence of states through periodic and event-triggered control for $p = 0.7$.

u_{total} represents the total number of time instants at which control input is actuated during the simulation period. According to Table 3.1, the proposed event-triggered control strategy significantly reduces the communication usage. The efficiency of event-triggering law may depend on different system dynamics, performance index, initial conditions, and other uncertainty profiles. Moreover the ratio $\frac{\|\Delta A\|}{\|A\|}$ is found to be 30.88% for $p = 0.7$. This shows that the nominal system is perturbed through a significant amount of uncertainty.

3.5 SUMMARY

A discrete-time periodic and aperiodic robust control law is proposed for an uncertain linear system. To design the robust control law, an equivalent optimal control problem is formulated for a virtual nominal system with a modified cost-functional. A virtual control input is defined to design the stabilizing controller gain along with the stability condition. This chapter also proposes an event-triggered-based control technique for NCS to achieve robustness. The event-condition and stability of mismatched uncertain system are derived using the ISS Lyapunov function. A comparative study between existing and proposed results is also reported. The next part of this book considers nonlinear system for design and analysis.

Section II

Control of Nonlinear Systems

4 Finite-time Event-triggered Control for a Class of Nonlinear Systems

After studying this chapter, one should be able to: discuss the issues and motivation for nonlinear event-triggered control problem; design an event-triggered control law for a class of nonlinear system; use state dependent Riccati equation to solve optimal control problem for a class of nonlinear systems; learn the techniques to convert event-triggered Hamilton-Jacobi Bellman (ET-HJB) equation to state dependent Riccati equation (SDRE); analyze the stability of a class nonlinear systems under event-triggered feedback; use the proposed control law in various application areas.

4.1 INTRODUCTION

In previous chapters, the event-triggered control law for linear system have been proposed. But controlling a nonlinear system with finite-time convergence of states under limited feedback information is a challenging research problem due to its time dependency of the solution. In [88] and [77], attempt has been made to solve an infinite-time optimal control problem for a linear event-triggered system. However extending these results for nonlinear systems with the finite-time convergence is not straightforward. In this chapter, an attempt is made to realize a finite-time event-triggered control law for a class of nonlinear system whose system dynamics can be represented in a state-dependent coefficient (SDC) form [28, 26, 90]. Applying the SDC form in system dynamics, the state and input functions are converted to a linear like structure. Figure 4.1 describes the overall block diagram of proposed control method. Here the system and controller are connected through a network. The state and input information are transmitted aperiodically to the controller and actuator respectively. To design a finite-time event-triggered control law, an event-triggered Hamilton-Jacobi-Bellman (ET-HJB) equation is considered. After simplification, the ET-HJB is converted into a state-dependent differential Riccati equation (SDRE). The results reported in [60] are used for initial simplification. Applying the frozen-time concept within two consecutive events, the SDRE reduces to a simple differential Riccati equation (DRE). The analytical solution of DRE is calculated using the solution of a differential Lyapunov equation (DLE). The solution of DLE is used to compute the control law at every event-triggering instant. This chapter makes the following contributions:

DOI: 10.1201/9781003229698-4

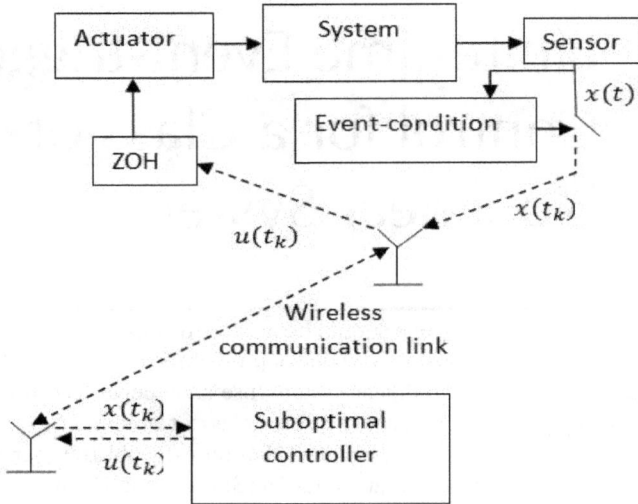

Figure 4.1 Block diagram of proposed event-triggered control technique. The dotted lines are used to represent the aperiodic transmission of data over the communication link.

(i) the proposed finite-time optimal control approach is applicable for a class of nonlinear system (whose system dynamics can be expressed in SDC form). To generate the control input, the ET-HJB is converted into an SDRE. The solution of SDRE is computed based on the solution of DLE. Using the analytical solution of SDRE for a final time t_f, the control input is computed at the every event-triggering instant.

(ii) An event-triggering condition for a nonlinear system is proposed by solving an SDRE equation. The event-triggered control law ensures ISS of the closed loop system. A comparative study between continuous and event-triggered method is carried out to highlight the contribution of this chapter over the existing literature.

(iii) An example with simulation and experimental results are used to validate the proposed control algorithm. The control law and triggering condition are implemented on a computing units general purpose PC and Intel Galileo board, respectively.

Organization

The chapter is organized as follows. Section 4.2 defines the research problem, which is addressed in this chapter. The main results of this chapter is reported in Section 4.3. This section also compares the proposed results with the existing work reported in [60]. The experimental and numerical validation of proposed control algorithm are reported in Section 4.4. The Section 4.5 summarizes the main contributions of the chapter.

4.2 PROBLEM FORMULATION

Consider a continuous-time input-affine nonlinear system with state-dependent coefficient as [28, 26, 90]

$$\dot{x}(t) = A(x)x(t) + B(x)u(t) \tag{4.1}$$

where state and input matrices are $A(x) \in \mathbb{R}^{n \times n}$ and $B(x) \in \mathbb{R}^n$. Here $x \in \mathbb{R}^n$ and $u \in \mathbb{R}$ represent the system state and input vector, respectively. To stabilize (4.1), a finite-time optimal control law is designed by minimizing the following quadratic cost-functional

$$J = \frac{1}{2} x_f^T F x_f + \frac{1}{2} \int_{t_0}^{t_f} \left(x^T Q x + u^T R u \right) dt \tag{4.2}$$

where matrices $F \geq 0$, $Q \geq 0$ and $R > 0$. The time-instant t_0 and t_f of (4.2) denote the initial and final time, respectively. To compute the optimal input for (4.1) with the cost-functional (4.2), the Hamiltonian H is defined as

$$H\left(x, u(t), \frac{\partial J}{\partial x}\right) = \frac{1}{2}\{x^T(t)Qx(t) + u^T(t)Ru(t)\} + \frac{\partial J}{\partial x}^T \left\{ A(x)x(t) + B(x)u(t) \right\}. \tag{4.3}$$

Using optimal control theory results [91], the optimal input $u^*(t)$ should minimize the H, which means

$$\frac{\partial H\left(x, u^*(t), \frac{\partial J^*(x,t)}{\partial x}\right)}{\partial u^*(t)} = 0. \tag{4.4}$$

The notation $J^*(x,t)$ denotes the optimal value of cost-functional J. After simplification, (4.4) reduces to

$$u^*(t) = K(x) = -R^{-1}B(x)^T \frac{\partial J^*(x,t)}{\partial x} \tag{4.5}$$

where $K(x)$ is a controller gain function. Also $u^*(t)$ satisfies the well-known HJB equation i.e.,

$$-\frac{\partial J^*(x,t)}{\partial t} = \frac{1}{2}\{x^T Q x + u^{*T} R u^*\} + \frac{\partial J^*(x,t)}{\partial x}^T \left\{ A(x)x(t) + B(x)u^*(t) \right\}. \tag{4.6}$$

Now to generate $u^*(t)$ from (4.5), the solution J^* of HJB equation (4.6) is essential. To avoid the difficulty of solving (4.6), Heydari et al. [60] proposed an approximation technique to generate the optimal control input $u^*(t)$ by converting (4.6) into a following SDRE [91]

$$\dot{P}(x,t) = P(x,t)A(x) + A^T P(x,t) - P(x,t)B(x)R^{-1}B^T(x)P(x,t) + Q. \qquad (4.7)$$

with the boundary condition

$$P(x,t_f) = F. \qquad (4.8)$$

The solution $P(x,t)$ of (4.7) is used to compute the approximate solution of (4.5) as

$$u(t) = -R^{-1}B(x)^T P(x,t)x(t). \qquad (4.9)$$

The feedback control input (4.9) ensures the stability of closed loop system (4.1).

The realization of control law (4.9) for NCS requires continuous transmission of x over the communication channel. This causes a huge communication overhead. Therefore to reduce the communication burden, the state and input information can be transmitted and actuated aperiodically. The aperiodic time-sequence of information transmission is denoted by t_k where $k \in \{1,2,3,4\cdots N\}$. Due to aperiodic nature of information transmission, the continuous control input $u(t)$ is converted to $u(t_k)$. Applying $u(t_k)$, the closed loop system (4.1) and HJB equation (4.6) reduce to

$$\dot{x} = A(x)x + B(x)u(t_k) \qquad (4.10)$$

$$-\frac{\partial J^*}{\partial t} = \frac{1}{2}\left\{x(t)^T Qx(t) + u^*(t_k)^T Ru^*(t_k)\right\} + \left(\frac{\partial J^*}{\partial x}\right)^T \left\{A(x)x(t) + B(x)u^*(t_k)\right\}. \qquad (4.11)$$

From [117] and [52], system (4.10) can be modeled as continuous time perturbed system by introducing an error variable $e(t)$, which is defined as

$$e(t) = x(t_k) - x(t), \ \forall t \in [t_k, t_{k+1}). \qquad (4.12)$$

At the event-triggering instant t_k, the numerical value of measurement error $e(t)$ is zero. Using $e(t)$, the event-triggered control input $u(t_k)$ reduces to

$$u(t_k) = K(x,e) = -R^{-1}B(x(t_k))^T \frac{\partial J^*(x(t),t)}{\partial x(t)}\bigg|_{t=t_k}, \forall t \in [t_k, t_{k+1}). \qquad (4.13)$$

For subsequent analysis, the following assumption is made.

Assumption 1 *The event-triggered closed loop system (4.10) is Lipschitz continuous with respect to state $x(t)$ and measurement error $e(t)$ that means*

$$\|A(x)x(t) + B(x)K(x,e)\| \le L_1 \|x(t)\| + L_2 \|e(t)\| \qquad (4.14)$$

where L_1 and L_2 are positive constants.

To generate (4.13), it is essential to solve or approximate the ET-HJB (4.11). To achieve this goal, several researchers have used neural network (NN) as an universal function approximater to estimate the optimal value function J^* [127, 110]. For estimation of J^*, they update the NN weight vector aperiodically to reduce the computation burden. Apart from NN-based approximation technique, Heydari et al. [60] have approximated the equation (4.6) as an SDRE for a class of nonlinear system to obtain a stabilizing control law (4.9). But with an event-triggered control input $u(t_k)$, the conversion of ET-HJB (4.11) to a SDRE is a non-trivial problem. To obtain an SDRE from (4.11), the conversion procedure and the numerical tools for solving SDRE are discussed in next section. For the stability analysis of the closed loop event-triggered system (4.10), the ISS property is adopted. The definition of ISS theory [71, 115, 93] has been introduced in Chapter 2.

Problem Statement: Convert the ET-HJB equation (4.11) into an SDRE and design a state feedback event-triggered control as

$$u(t_k) = K\{B(x(t_k)), P(x(t_k), t_k), x(t_k)\} \tag{4.15}$$

where vector $x(t_k)$ and matrix $P(x(t_k), t_k)$ are the aperiodic state and the positive definite solution of the proposed Riccati equation, respectively. Derive the control law (4.15) in order to ensure the ISS of closed loop system (4.10) for a given event-triggering rule.

4.3   MAIN RESULTS

The primary theoretical contributions of this chapter are divided into three parts and they are discussed in next two subsections. Firstly, the conversion of ET-HJB to SDRE is discussed and then using the solution of SDRE, the stability of closed loop system is ensured. Along with this stability result, a state dependent event-triggering rule is also defined. Finally, a numerical procedure to solve the proposed SDRE is reported. The following standard assumption [71, 75] is used to convert (4.11) into an SDRE.

Assumption 2 *For a positive constant L and measurement error $e(t)$, the event-triggered control $u(t_k)$ and continuous control $u(t)$ hold the following inequality*

$$\|u(t) - u(t_k)\| \le L\|e(t)\|, \ \forall t \in [t_k, t_{k+1}). \tag{4.16}$$

4.3.1   CONVERSION OF ET-HJB TO SDRE

The conversion processes of ET-HJB into an SDRE is stated next.

Theorem 10 *Suppose there exist positive scalars L and $\sigma \in (0,1)$ such that the Assumption 2 holds. For an event-triggering rule*

$$t_0 = 0, \ t_{k+1} = inf\{t \in \mathbb{R} | t > t_k \wedge (\mu\|x\|^2 - \|e\|^2 \le 0)\} \tag{4.17}$$

the ET-HJB (4.11) reduces to a following SDRE

$$-\dot{P}(x,t) \leq P(x,t)A(x) + A^T(x)P(x,t)$$
$$-P(x,t)B(x)B(x)P(x,t) + (\sigma+1)Q - \mho \qquad (4.18)$$

where scaling matrix $R = I$ and the boundary condition $P(x,t_f) = F$ are selected. The parameter μ and matrix $\mho$ are defined as

$$\mu = \frac{\sigma \lambda_{min}(Q)}{L^2} \qquad (4.19)$$

$$\mho = \left\{ \sum_{i=1}^{n} P_{x_i} z_i + \frac{1}{4}\left(\sum_{i=1}^{n}\sum_{j=1}^{n} P_{x_i} x (B(x)B(x)^T)_{ij} x^T P_{x_j}\right) + P(B(x)B(x)^T) \begin{bmatrix} x^T \frac{\partial P(x,t)}{\partial x_1} \\ \vdots \\ x^T \frac{\partial P(x,t)}{\partial x_n} \end{bmatrix} \right\} \qquad (4.20)$$

where scalar $z_i = b_i u(t_k)$ and b_i is the ith element of input matrix $B(x)$. The notation $[\,\cdot\,]_{ij}$ represents the ijth element of matrix $[\,\cdot\,]$. The partial derivative of a positive matrix $P(x,t)$ with respect to the individual state element $x_{i \in \{1 \leq i \leq n\}}$ is denoted by P_{x_i}.

Proof *Using $R = I$, the ET-HJB (4.11) reduces to*

$$-\frac{\partial J^*}{\partial t} = \frac{1}{2}(x^T Q x + u(t_k)^T u(t_k)) + \left(\frac{\partial J^*}{\partial x}\right)^T \left\{A(x)x(t) + B(x)u(t_k)\right\} \qquad (4.21)$$

which has an optimal cost-to-go

$$J^*(x,t) = x^T P(x,t)x. \qquad (4.22)$$

The partial derivative of $J^*(x,t)$ with respect to t and x are simplified as

$$\frac{\partial J^*}{\partial t} = \frac{1}{2}x^T P_t x \qquad (4.23)$$

$$\frac{\partial J^*}{\partial x} = Px + \Gamma \qquad (4.24)$$

where the matrices $P_t = \frac{\partial P}{\partial t}$ and

$$\Gamma = \frac{1}{2}\begin{bmatrix} x^T \frac{\partial P(x,t)}{\partial x_1} x \\ \vdots \\ x^T \frac{\partial P(x,t)}{\partial x_n} x \end{bmatrix}. \qquad (4.25)$$

Using (4.23) and (4.24), the ET-HJB (4.21) reduces to

$$-\frac{1}{2}x^T P_t x = \frac{1}{2}\{x(t)^T Qx(t) + u(t_k)^T u(t_k)\} + (Px + \Gamma)^T \{A(x)x + B(x)u(t_k)\}. \quad (4.26)$$

Further, (4.26) is simplified as

$$-\frac{1}{2}x^T P_t x - \Gamma^T \dot{x} = \frac{1}{2}\{x^T Qx + u(t_k)^T u(t_k)\} + x^T P(x,t)A(x)x + x^T P(x,t)B(x)u(t_k). \quad (4.27)$$

The above equation (4.27) can be written as

$$-\frac{1}{2}x^T \dot{P}(x,t)x = \frac{1}{2}\{x^T Qx + u(t_k)^T u(t_k)\} + x^T P(x,t)A(x)x + x^T P(x,t)B(x)u(t_k) \quad (4.28)$$

where the following auxiliary equations are used

$$\Gamma^T \dot{x} = \frac{1}{2}x^T \left(\sum_{i=1}^{n} P_{x_i}\dot{x}_i\right)x$$

$$-\frac{1}{2}x^T P_t x - \frac{1}{2}x^T \left(\sum_{i=1}^{n} P_{x_i}\dot{x}_i\right)x = -\frac{1}{2}x^T \dot{P}x.$$

From (4.5) and (4.24), the optimal input $u^*(t)$ can be written as

$$u^{*T}(t) = -(Px + \Gamma)^T B(x). \quad (4.29)$$

For simplicity, subsequently the optimal cost-functional J^*, the matrices $P(x,t)$, $A(x)$, $B(x)$, and input $u^*(t)$ are denoted by J, P, A, B, and u(t), respectively. Applying (4.29), the term $x^T PB(x)u(t_k)$ is simplified as

$$\begin{aligned} x^T PBu(t_k) &= (Px + \Gamma)^T Bu(t_k) - \Gamma^T Bu(t_k) \\ &= -u^T(t)u(t_k) - \Gamma^T Bu(t_k). \end{aligned} \quad (4.30)$$

Using (4.30), equation (4.28) reduces to following equality

$$-\frac{1}{2}x^T \dot{P}x = \frac{1}{2}\{x^T Qx + u(t_k)^T u(t_k)\} + x^T PAx + \frac{1}{2}(u(t) - u(t_k))^T(u(t) - u(t_k)) - \frac{1}{2}\{u(t)^T u(t) + u(t_k)^T u(t_k)\} - \Gamma^T Bu(t_k). \quad (4.31)$$

The event-triggering rule (4.17) is used to define an upper bound of (4.16). Now using (4.16) and (4.17), equality (4.31) reduces to following inequality

$$-\frac{1}{2}x^T \dot{P}x \leq \frac{(\sigma+1)}{2}x^T Qx + x^T PAx - \frac{1}{2}u^T u - \Gamma^T Bu(t_k). \quad (4.32)$$

Equation (4.29) is used to simplify the above inequality (4.32).

$$-x^T \dot{P}x \leq (\sigma+1)x^T Qx + x^T PAx + x^T A^T Px - x^T PBB^T Px - [2x^T PBB^T \Gamma$$
$$+ \Gamma^T BB^T \Gamma + 2\Gamma^T Bu(t_k)]. \tag{4.33}$$

To construct a Riccati like equation, it is essential to express the terms $2\Gamma^T Bu(t_k)$, $\Gamma^T B(x)B(x)^T \Gamma$ and $2x^T PB(x)B(x)^T \Gamma$ in the quadratic form. To obtain that, following simplifications are adopted:

> *The event-triggered input $u(t_k)$ is constant within the inter-event time. The input matrix $B(x) = \begin{bmatrix} b_1(x) & b_2(x) & \cdots & b_n(x) \end{bmatrix}^T$ depends on system's sate x. Therefore, the term $B(x)u(t_k)$ can be represented by another different state dependent variable $Z(x)$ and it is defined as follows*

$$Z(x) = \begin{bmatrix} z_1(x) & z_2(x) & \cdots & z_n(x) \end{bmatrix}^T \tag{4.34}$$

where $z_i(x) = b_i(x)u(t_k)$ for $i = 1, 2 \ldots n$. Using (4.34) and (4.25), following equalities are achieved

$$2\Gamma^T Bu(t_k) = x^T (\sum_{i=1}^{n} P_{x_i} z_i)x, \tag{4.35}$$

$$\Gamma^T BB^T \Gamma = \frac{1}{4}x^T (\sum_{i=1}^{n} \sum_{j=1}^{n} P_{x_i} x(BB^T)_{ij} x^T P_{x_j})x. \tag{4.36}$$

Similarly, the term $2x^T PBB^T \Gamma$ is reduces to

$$2x^T PBB^T \Gamma = x^T PBB^T \begin{bmatrix} x^T \frac{\partial P(x,t)}{\partial x_1} \\ \cdots \\ x^T \frac{\partial P(x,t)}{\partial x_n} \end{bmatrix} x. \tag{4.37}$$

Using (4.35), (4.36), and (4.37), the equation (4.33) is simplified as

$$-x^T \dot{P}x \leq x^T PAx + x^T A^T Px - x^T PBB^T Px + (\sigma+1)x^T Qx - x^T \mho x \tag{4.38}$$

where matrix $\mho$ is defined in (4.20). Now rewriting the inequality (4.38), (4.18) is achieved for a final condition $P(x,t_f) = F$. The proof is completed.

Remark 15 *For an equality relation, the equation (4.18) is converted to an approximated Riccati equation by neglecting the term $\mho$ from (4.18). The expression of approximated Riccati equation is*

$$-\dot{P}(x,t) = (\sigma+1)Q + P(x,t)A(x) + A^T(x)P(x,t) - P(x,t)B(x)B(x)^T P(x,t) \tag{4.39}$$

with a boundary condition $P(x,t_f) = F$. The constant matrix $(\sigma+1)Q > 0$ in (4.39) is changed from (4.7) due to the aperiodic update of control inputs. The

> *solution P of (4.39) is used to compute the event-triggered control input (4.15). The approximate control input will not be an optimal one, but it holds the ISS property for (4.10).*

The following theorem ensures the ISS of (4.10) for the event-triggering law (4.17).

Theorem 11 *Suppose there exists a positive definite solution $P(x,t)$ of (4.39) which is aperiodically computed for an event-triggering rule (4.17). The solution $P(x,t)\Big|_{t=t_k}$ and aperiodic state information $x(t_k)$ generate the event-triggered control input*

$$u(t_k) = -B(x(t_k))^T P(x(t_k), t_k) x(t_k) \tag{4.40}$$

which is actuated at the system end based on the event-triggering rule (4.17). The control input (4.40) ensures the ISS of closed loop system (4.10).

Proof *Defining an ISS Lyapunov function $V(x) = x^T P(x,t)x$, the time derivative of $V(x)$ along the solution of (4.10) is*

$$\dot{V}(x) = -(\sigma+1)x^T Qx + (u(t) - u(t_k))^T (u(t) - u(t_k)) - u(t_k)^T u(t_k). \tag{4.41}$$

After further simplification, the upper bound of (4.41) is

$$\dot{V}(x) \leq -(\sigma+1)x^T Qx + (u(t) - u(t_k))^T (u(t) - u(t_k)). \tag{4.42}$$

Using Assumption 2 in (4.42), following equation is obtained

$$\dot{V}(x) \leq -\sigma x^T Qx + L^2 \|e\|^2. \tag{4.43}$$

Using Definition 10, (4.43) ensures the ISS of (4.10) for an input (4.40) with the event-triggering rule (4.17).

Remark 16 *From (4.36) and (4.20) it is possible to show that there exist two constants c_1 and c_2 such that the Γ and $\mho$ are bounded by the following equations*

$$\Gamma \leq c_1 \|x\|^2, \tag{4.44}$$
$$\mho \leq c_2 \|x\|^2. \tag{4.45}$$

Therefore as $x(t) \to 0$ for $t \to \infty$, the terms Γ and $\mho$ also approach toward zero. This ensures that the control input (4.40) also converges to the suboptimal solution [60, 90].

Remark 17 *In SDC form of (4.10), the state and input matrices depend on state information. But in controller, continuous state information is not available due to communication constraint. To resolve this problem we have adopted frozen-time concept from [105] in order to solve (4.39). It is assuming that the state and input matrices are remain constant in-between two consecutive events. This also helps to solve the Riccati equation in a frozen-time manner.*

Remark 18 *In event-triggered control it is essential to prove that the inter-event time* $\tau = (t_{k+1} - t_k)$ *is always positive i.e.,* $\tau > 0$. *This constraint is imposed to avoid the Zeno behavior [69] in system dynamics. Now for a system (4.10) whose initial condition* $x(0)$ *remains in a compact set* $S \subseteq \mathbb{R}^n$ *(i.e.,* $x(0) \in S$), *there exists a lower bound* $\tau \in \mathbb{R}^+$ *for the event-triggering rule (4.17), which satisfy* $t_{k+1} - t_k \geq \tau$, $\forall k \in \mathbb{N}$ *[117].*

4.3.2 NUMERICAL SOLUTION OF SDRE

As per the Remark 17, the derived SDRE (4.39) can be considered as following DRE within the two consecutive events [105]

$$-\dot{P}(t, x(t_k)) = P(t)A(x(t_k)) + A^T(x(t_k))P(x(t_k), t)$$
$$-P(x(t_k), t)B(x(t_k))B(x(t_k))^T P(x(t_k), t) + (\sigma + 1)Q. \quad (4.46)$$

The solution procedure of (4.46) is discussed next [96]:

Compute the steady state value (P_{ss}) of (4.46) , by solving the following ARE:

$$P_{ss}A + A^T P_{ss} - P_{ss}BB^T P_{ss} + (\alpha + 1)Q = 0. \quad (4.47)$$

Subtracting (4.46) from (4.47) following is obtained

$$-\dot{P}(t) = (P - P_{ss})A + A^T(P - P_{ss}) - P(t)SP(t) + P_{ss}^T SP_{ss}. \quad (4.48)$$

where matrix $S = B(x(t_k))B(x(t_k))^T$.
Using a change of variable $P_0(t) = (P(t) - P_{ss})^{-1}$, the equation (4.48) is converted into the following differential Lyapunov equation [92]

$$\dot{P}_0(t) = A_{cl}P_0(t) + P_0(t)A_{cl}^T - S \quad (4.49)$$

with the final condition $P_0(t_f) = (F - P_{ss})^{-1}$ and $A_{cl} = A(x(t_k)) - SP_{ss}$.
Compute the solution of algebraic Lyapunov equation

$$A_{cl}D_0 + D_0A_{cl}^T - S = 0. \quad (4.50)$$

The solution of (4.49) is

$$P_0(t) = e^{A_{cl}(t-t_f)}(P_0(t_f) - D_0)e^{A_{cl}^T(t-t_f)} + A_0. \tag{4.51}$$

The solution of original SDRE (4.46) is

$$P(t) = P_{ss} + P_0(t)^{-1}. \tag{4.52}$$

The expression (4.52) is used to compute (4.40).

Remark 19 *Here we compare the main contributions of this chapter with the existing research work [60]. In [60], the HJB equation has been approximated as an SDRE. To approximate the HJB equation into an SDRE, the primary assumption in the work is that the state and input information are continuously available to the controller and system end, respectively. In general, this assumption does not hold for event-triggered control technique due to asynchronous availability of state and input information. In [126], K. G. Vamvoudakis has named the equation (4.11) as ET-HJB equation for the presence of aperiodic control input $u(t_k)$. This chapter proposes a procedure to convert ET-HJB equation into an SDRE, which helps to solve the nonlinear optimal control problem for a class of systems with limited feedback information. The presence of aperiodic input $u(t_k)$ in (4.11), complicates the conversion processes of ET-HJB equation into an SDRE. The detailed steps of conversion processes are described in this chapter [refer Theorem 10 and its proof]. Due to limited availability of state and control input, we obtain a state-dependent differential Riccati like inequality (4.18). The inequality (4.18) is converted into a Riccati equation (4.39), using an equality relation. The positive definite solution P of (4.39) is evaluated at every event-triggering instant (t_k), to compute the aperiodic control input $u(t_k)(= -B(x(t_k))^T P(x(t_k), t_k)x(t_k))$. To handle the aperiodic actuation of control input $u(t_k)$, the constant matrix Q of (4.39) is scaled by $(1+\sigma)$ times compared to reported work [60]. In [60], the solution P of Riccati equation [equation (15) in [60]] has been computed continuously, but in this chapter a frozen-time approach is adopted to solve (4.39) aperiodically.*

4.4 RESULTS

In this section a benchmark example, the regulation problem of *Van der Pol's Oscillator* system is considered to prove the efficacy of the proposed control algorithm. Validation of the control problem is solved both numerically and experimentally.

Simulation

The *Van der Pol's Oscillator* model is described as

$$\begin{bmatrix} \dot{x}_1 \\ \dot{x}_2 \end{bmatrix} = \begin{bmatrix} 0 & x_2 \\ -1 & 1-x_1^2 \end{bmatrix} x + \begin{bmatrix} 0 \\ 1 \end{bmatrix} u \tag{4.53}$$

Table 4.1

Comparative Results of Event-triggered and Continuous Control Approach

Control mechanism		Performance for $t_f = 12$ sec.		
		$\tau_{max(s)}$	$\tau_{min(s)}$	u_{total}
Simulation Results	Continuous control	0.01	0.01	1200
	Event-triggered control	0.7149	0.01	144
Experimental Results	Continuous control	0.046	0.046	307
	Event-triggered control	3.169	0.046	61

For simulation, matrices $Q = I$, $R = I$, and $F = 10I$ are selected. To realize the event-triggering law (4.40), the scalar L and design parameter σ are considered as 0.82 and 0.1, respectively. The simulation is carried out in MATLAB with a final time $(t_f =)$ 2 and 12 sec. The initial states are selected as $\begin{bmatrix} 0.5 & 0.5 \end{bmatrix}^T$. Figures 4.2(b)–4.2(c) show the time evolution of state x for different final time t_f. The black and blue lines are used to represent the state trajectories of the system using the event-triggered and continuous control input, respectively. From the Figures 4.2(b)–4.2(c), it is observed that the proposed event-based control input helps to bring the system states near to the equilibrium point for a given finite-time. The aperiodic variation of control input $u(t_k)$ is shown in Figure 4.2(d). The numerical results show the ISS of closed loop system (4.10) and ensure the finite-time convergence of state trajectories for an aperiodic update of control input $u(t_k)$. From Table 4.1 it is observed the total number of input computation for continuous approach is comparatively higher than the event-triggered one. The minimum and maximum bound of inter-event time [τ_{min} and τ_{max}, respectively] of proposed event-triggering rule are shown in Table 4.1. The notation u_{total} represents the number of input computation in total runtime.

Experiment

To validate the proposed control algorithm on a real physical system, an experimental set-up is created. A *Van der Pol's Oscillator* circuit is developed as a physical system. The detailed circuit diagram of developed system is shown in Fig. 4.3. In this experiment, our goal is to regulate the oscillatory states to an equilibrium point zero with a minimal feedback information. To achieve this goal, an event-triggered nonlinear control law (4.40) is designed and it is realized on a computing unit. For a verification of derived event-triggering condition, an Intel Galileo board is considered which is connected with the physical system. This board collects the states of oscillator circuit as mentioned in (4.53) and computes the event-triggering condition based on the equation (4.17). The board is equipped with wireless XBee module, which acts as a transceiver at the system end. A general purpose PC equipped with another XBee module acts as a controller. The system (4.53) is used to compute the control law in MATLAB for a final time $t_f = 12$ sec. To compute the control law

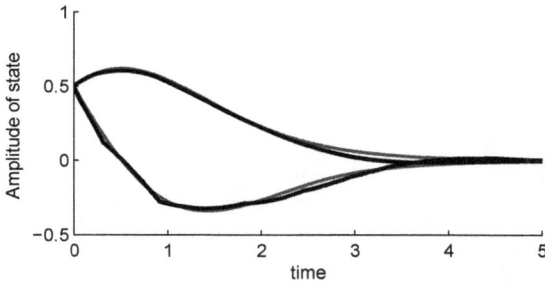

(a) Convergence of system states with $u(t_k)$ for $t_f = 5sec$.

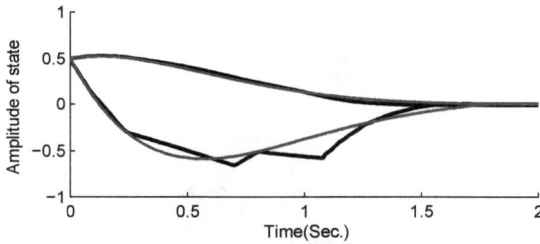

(b) Convergence of system states with $u(t_k)$ for $t_f = 2sec$.

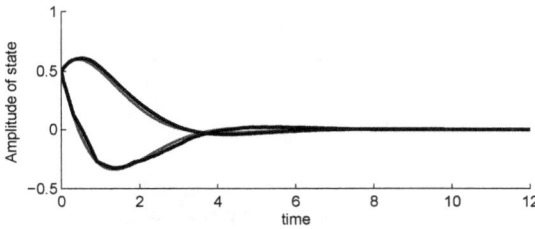

(c) Convergence of system states with $u(t_k)$ for $t_f = 5sec$.

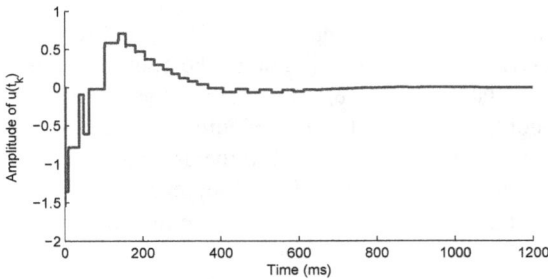

(d) Variation of control input $u(t_k)$ with respect to the time t for $t_f = 12sec$.

Figure 4.2 Simulation results of continuous and event-triggered control

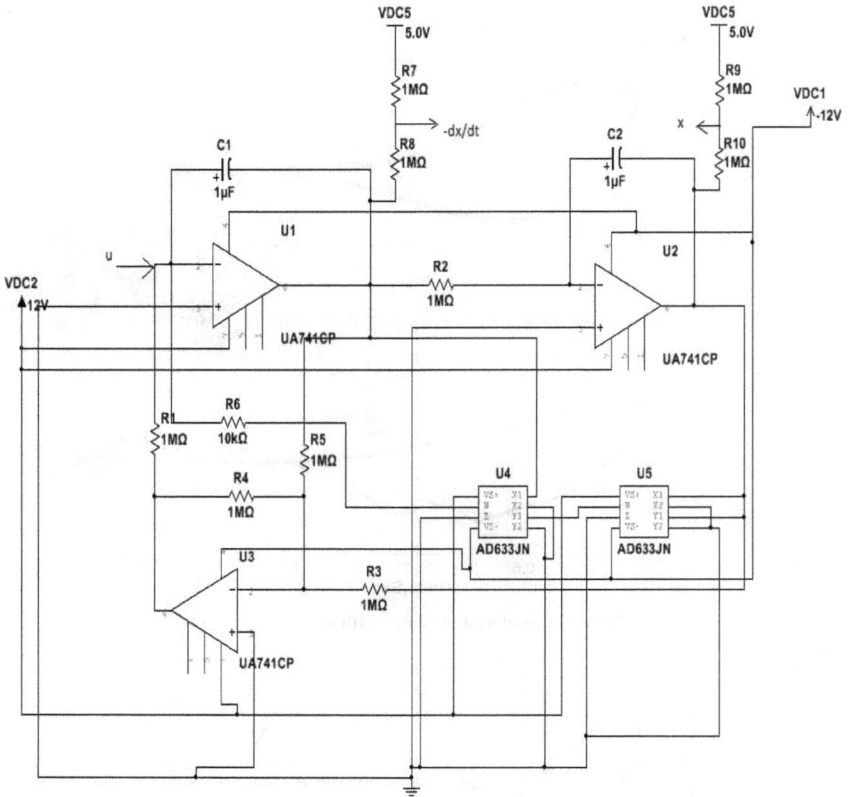

Figure 4.3 Electronic circuit diagram

and event-triggering condition, the matrices Q, R, F, and design parameters σ, L are selected as described earlier in simulation. The control inputs are updated aperiodically at the oscillator end through the XBee transmitter. The interconnection between each subcomponents are shown in Figure 4.4. The circuit diagram with the necessary components which are used to built the oscillator circuit are mentioned in Fig. 4.3. Figures 4.5(a) and 4.5(b) show the convergence of system states and control input for the event-triggered approach. To compare the efficiency of proposed control algorithm over the conventional continuous approach, Table 4.1 is introduced. The Table 4.1 shows the total number of transmission requirement for continuous and event-triggered case.

Physical System
(Van der Pol oscillator)

Intel Galileo Board
with Xbee module

[Event monitoring unit]

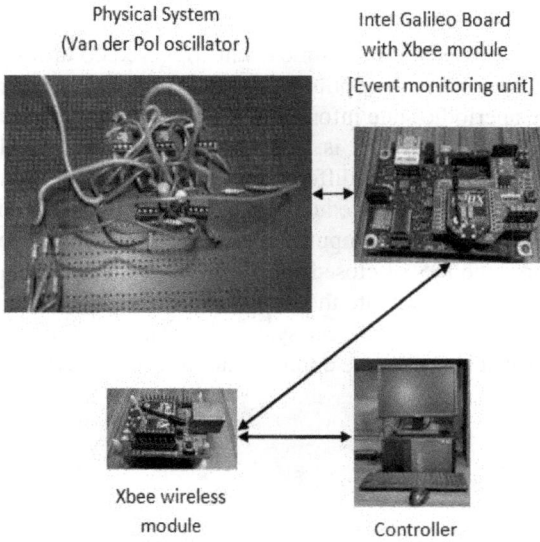

Xbee wireless
module

Controller

Figure 4.4 Experimental setup for event-triggered control

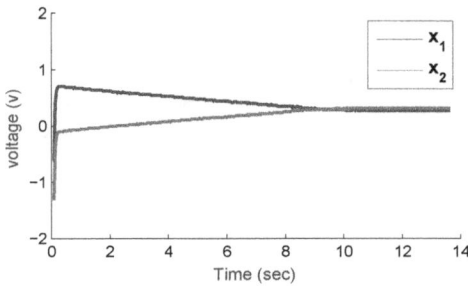

(a) Convergence of system states with $u(t_k)$.

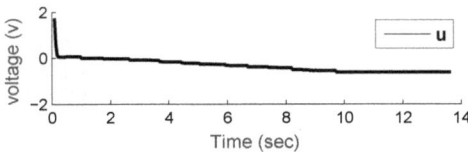

(b) Variation of $u(t_k)$ with respect to time (t).

Figure 4.5 Experimental results of the proposed event-triggered control

4.5 SUMMARY

Through this chapter, a novel finite-time event-triggered control law is proposed for a class of nonlinear system. An optimal control framework is adopted to design the control law using aperiodic state information. The proposed control law is computed by solving an SDRE equation. It is possible by converting the HJB equation into an SDDRE which is an ordinary differential equation. A DLE is solved to compute the solution of SDDRE. For the reduction of computation and communication burden, the solution of SDDRE is computed aperiodically. The proposed event-triggered control law ensures the ISS of closed-loop system. Both numerical and experimental results are provided to validate the proposed control algorithm. A robust control problem for a discrete-time nonlinear system is reported in next chapter where the control law is derived based on the approximate solution of HJB equation.

5 Robust Stabilization of Discrete-time Mismatched Nonlinear System

After studying this chapter, one should be able to: design an optimal control law for a discrete-time nonlinear system using the approximated solution of the discrete-time General Hamilton-Jacobi-Bellman (DT-GHJB) equation; design a robust control law for discrete-time nonlinear system using proposed optimal control framework; learn neural network (NN)-based approximation technique for solving DT-GHJB equation; prove the stability of discrete-time nonlinear system using approximate control input under presence of model uncertainties; validate the proposed control law numerically.

5.1 INTRODUCTION

In Chapter 4, an exact system model has been considered to derive a feedback control law for a nonlinear system. However, the need of an exact system model to design a feedback control law is the primary shortcoming of the classical feedback control technique. An uncertain system model is a more realistic representation and has far greater significance over the exact system model. To deal with parametric uncertainty in the system model, F. Lin and D. Wang et al. have proposed a continuous-time robust control technique for both linear and nonlinear system [79, 80, 82, 133, 131]. In both the cases, they have formulated an equivalent optimal control problem to derive the proposed robust control input. The optimal control problem is solved based on the nominal dynamics by minimizing a quadratic cost-functional with the knowledge of uncertainty bound. In [132], the results reported in [79, 80] are extended for a discrete-time nonlinear system. The robust control law is realized with an assumption that the physical system is affected by matched uncertainty. The controller design with mismatched uncertainty in system dynamics is a challenging research problem.

In this chapter, a discrete-time robust control technique for an uncertain nonlinear system is proposed. The system is primarily affected by mismatched uncertainty due to bounded parametric variation. To stabilize such systems, a robust control law is derived by solving a nonlinear optimal control problem for a nominal virtual system with a cost-functional. To solve the nonlinear optimal control problem, the solution of a discrete-time General Hamilton-Jacobi-Bellman (DT-GHJB) equation is

DOI: 10.1201/9781003229698-5

computed using a NN-based approximation technique. Based on the approximate so-lution of DT-GHJB, the cost-functional and control inputs are estimated. The block diagram representation of proposed control approach is shown in Fig. 5.1. The ap-proximate control inputs are used to derive the stability results for the uncertain sys-tem. Finally, numerical results are stated to prove the efficacy of the proposed control algorithm.

The key contributions of this chapter are:

(i) a robust control algorithm is proposed for a discrete time nonlinear system with mismatched uncertainty. The robust control law is derived by formulating an equivalent optimal control problem for a nominal virtual system with a quadratic cost-functional. The virtual dynamics have two control inputs u and v. The concept of virtual input v is used to derive the existence of stabiliz-ing input u. The virtual input v helps to tackle the mismatched uncertainty. The proposed robust control law ensures asymptotic convergence of uncertain closed-loop system.

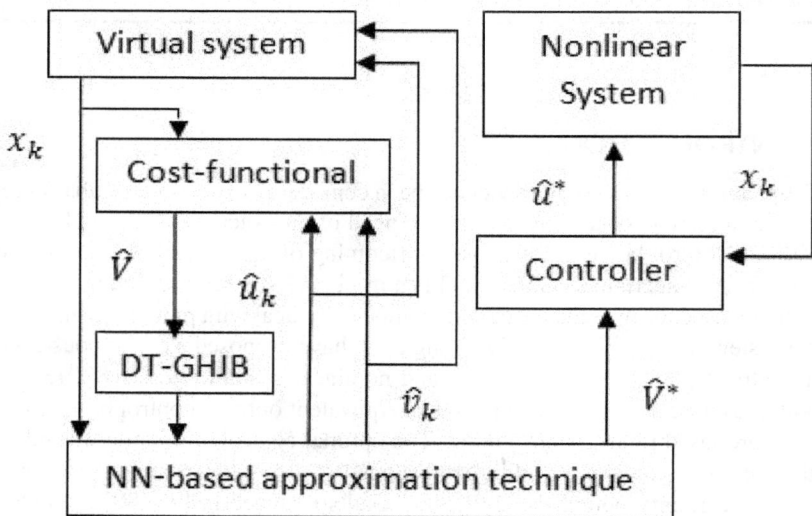

Figure 5.1 The block diagram of the proposed discrete-time robust control tech-nique is shown in this figure. Here notations x_k, $\hat{u}_k$, and $\hat{v}_k$ represent the system's state and two estimated control inputs, respectively. Using NN-based approximation technique, the estimated cost-functional $\hat{V}$ converges to its optimal cost $\hat{V}^*$. Using $\hat{V}^*$, the optimal inputs $\hat{u}^*$ and $\hat{v}^*$ are computed. Input $\hat{u}^*$ is applied to the nonlinear uncertain system to solve the robust control problem.

(ii) This chapter also shows that some of the existing results [132] of matched system are special cases of the proposed results.

(iii) To compute the optimal control input for a nonlinear system, the solution of DT-GHJB equation is approximated through a NN implementation. The approximate inputs ensure the asymptotic convergence of state trajectories. The convergence of both the NN weight and cost-functional are also shown through the simulation results.

To prove the theoretical results, following definition [112, 17] is used in this chapter.

Definition 11 *Consider a nonlinear discrete-time system as*

$$x_{k+1} = f(x_k) + g(x_k)u_k(x_k) \tag{5.1}$$

where $x_k \in \mathbb{R}^n$ and $u_k \in \mathbb{R}^m$ are system state and input vector, respectively. The functions $f(x_k)$ and $g(x_k)$ are continuous nonlinear functions and $f(x_k) + g(x_k)u_k(x_k)$ is Lipschitz continuous on a set Ω including the origin. The control input $u_k(x_k)$ ensures the asymptotic convergence of closed loop system (5.1), $\forall x_k \in \Omega$. Let Ω_u is a set of admissible control inputs and input u_k minimizes the cost-functional

$$J_k = \frac{1}{2}\sum_{k=0}^{\infty}\left\{x_k^T Q x_k + u_k^T R u_k\right\}. \tag{5.2}$$

Then, the control input u_k is considered as admissible (i.e., $u_k \in \Omega_u$) with-respect to its state penalty function $x_k^T Q x_k$ and control energy penalty function $u_k^T R u_k$, $\forall x_k \in \Omega$, if the following conditions hold:

(i) *$\forall x_k \in \Omega$, input $u(x_k)$ is continuous.*

(ii) *$u_k(0) = 0$.*

(iii) *u_k must stabilize (5.1) for $\forall x_k \in \Omega$.*

(iv) *$\sum_{k=0}^{\infty}(x_k^T Q x_k + u_k^T R u_k) \leq \infty$, $\forall x_0 \in \Omega$.*

The chapter is organized as follows. The Section 5.2 proposes an optimal control framework to solve a robust stabilization problem for a discrete-time mismatched nonlinear system. To compute the nonlinear optimal control input, this chapter considers a NN-based technique to approximate the solution of DT-GHJB. This section also compares the proposed results with the existing work reported in [132, 25]. The numerical validation of proposed control algorithm is shown in Section 5.3. A summary of this chapter is reported in Section 5.4.

5.2 ROBUST CONTROL DESIGN

A discrete-time uncertain nonlinear system is described by the state equation in the form

$$x_{k+1} = f(x_k) + g(x_k)u_k + d(x_k) \qquad (5.3)$$

where $x_k \in \mathbb{R}^n$ is the state, $u_k \in \mathbb{R}^m$ is the periodic control input, and $f \in \mathbb{R}^n, g \in \mathbb{R}^{n \times m}$ are nonlinear functions. It is assumed that (5.3) is Lipschitz continuous on a compact set $\Omega \in \mathbb{R}^n$ and origin is the equilibrium point i.e., $f(0) = 0$ and $g(0) = 0$. The unknown function $d(x_k) \in \mathbb{R}^n$ is used to represent the system uncertainty and it is always upper bounded by a known function $d_{max}(x_k)$, that is

$$\|d(x_k)\| \leq d_{max}(x_k), \forall k. \qquad (5.4)$$

Generally, system uncertainties are classified as matched and mismatched uncertainty and they are defined as follows [100].

Definition 12 *System (5.3) will be referred as matched uncertain system if the uncertainty $d(x_k)$ satisfies the following*

$$
\begin{aligned}
d(x_k) &= g(x_k)\phi(x_k) & (5.5)\\
\|\phi(x_k)\| &\leq U_{matched}, \forall k & (5.6)
\end{aligned}
$$

where $\phi(x_k)$ is the unknown function and $U_{matched}$ is the upper bound of $\|\phi(x_k)\|$. In other words, $d(x_k)$ is in the range space of $g(x_k)$.

Definition 13 *System (5.3) has mismatched uncertainty if the uncertain component $d(x_k)$ is not in the range space of input matrix $g(x_k)$.*

For the simplification, uncertainty can be decomposed into matched and mismatched component as follows

$$d(x_k) = g(x_k)g(x_k)^+ S\phi(x_k) + (I - g(x_k)g(x_k)^+) S\phi(x_k), \forall k \qquad (5.7)$$

where $g(x_k)g(x_k)^+ S\phi(x_k)$ and $(I - g(x_k)g(x_k)^+)S\phi(x_k)$ are the matched and mismatched components, respectively. The matrix $g(x_k)^+ = (g^T(x_k)g(x_k))^{-1}g(x_k)^T$ denotes the left pseudo inverse of matrix $g(x_k)$ [61] and S is a scaling matrix where $S \neq g(x_k)$. For a matrix $S = g(x_k)$, the uncertainty (5.7) reduces to a matched one as defined in (5.5). The decomposition of uncertainty into a matched and mismatched components will be used to define a nominal virtual system for (5.3) and is discussed in the subsequent subsection.

Problem Statement: Design a state feedback control law $u_k = K(x_k)$, to stabilize the discrete-time uncertain nonlinear system (5.3), such that the closed-loop system is asymptotically stable in the presence of uncertainty (5.7).

Proposed solution: This problem is solved in two steps. First, the controller is designed by adopting nonlinear optimal control theory and then an algorithm is used to approximate the solution of DT-GHJB equation. The approximate solution of DT-GHJB equation is used to compute the stabilizing and virtual control inputs u_k and v_k, respectively.

Robust control problem

Design a state feedback control law $u_k = K(x_k)$ such that the uncertain closed-loop system (5.3) is asymptotically stable $\forall \| d(x_k) \| \leq d_{max}(x_k)$. In order to stabilize (5.3), the robust control law u_k is designed using the results of optimal control theory.

Optimal control approach

The key idea is to design a discrete-time nonlinear optimal control law for virtual nominal system by minimizing a cost-functional J, which depends on the upper-bound of system uncertainty. An extra term $(I - g(x_k)g(x_k)^+)Sv(k)$ is added with the nominal dynamics of (5.3) to define a virtual system (5.8). The derived optimal input for virtual system is shown to be a robust input for original uncertain system. The virtual dynamics and cost-functional for solving robust control problem are given below:

$$x_{k+1} = f(x_k) + g(x_k)u_k + M(x_k)v_k \tag{5.8}$$

$$J_k = \frac{1}{2}\sum_{k=0}^{\infty}\left\{ d_{max}^2(x_k) + v_{max}^2(x_k) + x_k^T Q x_k + \begin{bmatrix} u_k^T & v_k^T \end{bmatrix}\begin{bmatrix} R_1 & 0 \\ 0 & R_2 \end{bmatrix}\begin{bmatrix} u_k \\ v_k \end{bmatrix}\right\} \tag{5.9}$$

where matrices $M = (I - g(x_k)g(x_k)^+)S$, $Q \geq 0$, $R_1 > 0$ and $R_2 > 0$. Here v_{max} is a scalar.

Inspired by the results reported in [5, 25], the discrete-time HJB (DT-HJB) equation for (5.8) with the optimal cost-functional J_k^* of (5.9) is

$$J_k^* = \min_{u(x_k),v(x_k)} \frac{1}{2}\left\{ d_{max}^2(x_k) + v_{max}^2(x_k) + x_k^T Q x_k + \begin{bmatrix} u_k^T & v_k^T \end{bmatrix}\begin{bmatrix} R_1 & 0 \\ 0 & R_2 \end{bmatrix}\begin{bmatrix} u_k \\ v_k \end{bmatrix}\right\} + J_{k+1}^* \tag{5.10}$$

Using (5.10), the optimal control input for (5.8) is

$$\begin{bmatrix} u_k^* \\ v_k^* \end{bmatrix} = \begin{bmatrix} R_1^{-1}g(x_k)^T \frac{\partial J_{k+1}^*}{\partial x_k} \\ R_2^{-1}M(x_k)^T \frac{\partial J_{k+1}^*}{\partial x_k} \end{bmatrix}. \tag{5.11}$$

Let $V(x_k)$ be a positive definite continuously differentiable function, which satisfies $V(x_0) = J(x_0, u(x_0))$. Applying Taylor series expansion of the cost-functional, the DT-HJB (5.10) reduces to discrete-time general HJB (DT-GHJB) as in [25]

$$d_{max}^2(x_k) + v_{max}^2(x_k) + x_k^T Q x_k + u_k^T R_1 u_k + v_k^T R_2 v_k +$$
$$\nabla V^T[x_{k+1} - x_k] + \frac{1}{2}[x_{k+1} - x_k]^T \nabla^2 V[x_{k+1} - x_k] = 0 \tag{5.12}$$

where $\nabla^2 V = \begin{bmatrix} \dfrac{\partial^2 V(x_k)}{\partial x_{k_1}^2} & \dfrac{\partial^2 V(x_k)}{\partial x_{k_1}\partial x_{k_2}} & \cdots & \dfrac{\partial^2 V(x_k)}{\partial x_{k_1}\partial x_{k_n}} \\ \dfrac{\partial^2 V(x_k)}{\partial x_{k_2}\partial x_{k_1}} & \dfrac{\partial^2 V(x_k)}{\partial x_{k_2}^2} & \cdots & \dfrac{\partial^2 V(x_k)}{\partial x_{k_2}\partial x_{k_n}} \\ \vdots & \vdots & \ddots & \vdots \\ \dfrac{\partial^2 V(x_k)}{\partial x_{k_n}\partial x_{k_1}} & \dfrac{\partial^2 V(x_k)}{\partial x_{k_n}\partial x_{k_2}} & \cdots & \dfrac{\partial^2 V(x_k)}{\partial x_{k_n}^2} \end{bmatrix}$ and $\nabla V = \frac{\partial V(x_k)}{\partial x_k}$. The

notation $x_{k_{i\in 1\cdots n}}$ represents the ith element of state vector x_k. The Hermitian matrix $\nabla^2 V$ is positive-definite $\forall\, x_k \in \Omega$. In the Taylor series expansion, the third- and
higher-order terms are dropped to make it computationally feasible. This is made
possible by adopting the small gain perturbation assumption around the equilibrium
point. Using (5.9) and (5.12), it can be proved that there exists a control input $u(x_k)$
which satisfies the equality $V(x_k) = J(x_k, u(x_k))$ [112]. Now according to optimal
control theory [91], the optimal inputs u_k^* and v_k^* satisfy the DT-GHJB and also minimize the following Hamiltonian:

$$H(x_k, u_k, v_k, \nabla V) = d_{max}^2(x_k) + v_{max}^2(x_k) + x_k^T Q x_k + u_k^T R_1 u_k + v_k^T R_2 v_k + \nabla V_k^T [x_{k+1} - x_k]$$
$$+ \frac{1}{2}[x_{k+1} - x_k]^T \nabla^2 V_k [x_{k+1} - x_k]. \tag{5.13}$$

That means $\frac{\partial H}{\partial u^*} = 0$ and $\frac{\partial H}{\partial v^*} = 0$, which correspond to

$$g^T \nabla^2 V^*(f + gu^* + Mv^* - x) + (2R_1 u^* + g^T \nabla V^*) = 0, \tag{5.14}$$
$$M^T \nabla^2 V^*(f + gu^* + Mv^* - x) + (2R_2 v^* + M^T \nabla V^*) = 0. \tag{5.15}$$

The scalar $V^*(x_k)$ is the optimal value of $V(x_k)$ and it satisfies equation (5.12). After
simplification, from (5.14) and (5.15), the optimal inputs are

$$\begin{bmatrix} u_k^* \\ v_k^* \end{bmatrix} = -\begin{bmatrix} (2R_1 + g^T \nabla^2 V^* g) & g^T \nabla^2 V^* M \\ M^T \nabla^2 V^* g & (2R_2 + M^T \nabla^2 V^* M) \end{bmatrix}^{-1} \begin{bmatrix} g^T(\nabla V^* + \nabla^2 V^*(f - x)) \\ M^T(\nabla V^* + \nabla^2 V^*(f - x)) \end{bmatrix}. \tag{5.16}$$

The control inputs (5.16) ensure the stability of (5.8). This information is stated
through the following Lemma.

Lemma 6 *Suppose there exists a Lyapunov function $V(x_k)$ for (5.8) and DT-
GHJB (5.12) is satisfied. Then, the optimal inputs u_k^* and v_k^* defined in (5.16)
ensure the asymptotic convergence of virtual nominal system (5.8).*

Proof Let $V(x_k)$ is the Lyapunov function for (5.8). Using (5.12), the $\Delta V(x_k) = V_{k+1} - V_k$ reduces to

$$\Delta V = -(x_k^T Q x_k + d_{max}^2(x_k) + v_{max}^2(x_k) + u_k^{*T} R_1 u_k^* + v_k^{*T} R_2 v_k^*). \tag{5.17}$$

*The negative-definiteness of ΔV along the solution of (5.8) proves the asymptotic
stability of (5.8) using the control inputs (5.16).*

In order to ensure the invertibility of matrix in (5.16), following remark is introduced.

Remark 20 *A block matrix* $\begin{bmatrix} A & B \\ \hline B^T & C \end{bmatrix}$ *is invertible if the following conditions are satisfied [65]:*

i) $det(A) \neq 0$,

ii) $det(C - BA^{-1}B^T) \neq 0$

Now, the control inputs (5.16) can be computed if the matrix

$$\begin{bmatrix} A & B \\ \hline B^T & C \end{bmatrix} = \begin{bmatrix} 2R_1 + g^T \nabla^2 V^* g & g^T \nabla^2 V^* M \\ \hline M^T \nabla^2 V^* g & 2R_2 + M^T \nabla^2 V^* M \end{bmatrix}$$

is invertible. Here R_1, R_2 and $\nabla^{*2}V$ are the positive definite matrices. So the sub-matrix $A(= 2R_1 + g^T \nabla^2 V^* g)$ is positive definite as $(2R_1 + g^T \nabla^2 V^* g) > 0$ and hence $det(A) \neq 0$. Now a suitable selection of design matrices R_1 and R_2 helps to satisfy condition (ii).

The realization of optimal control inputs (5.16) depends on the solution of DT-GHJB (5.12). In the next section, a brief description of NN-based approximation technique is discussed to achieve the estimated solution of (5.12), which helps to design the optimal inputs (5.16).

5.2.1 NN-BASED APPROXIMATION USING LEAST SQUARES APPROACH

NN has universal function approximation property. Using this property, several researchers have used NN to approximate the solution of HJB or GHJB as reported in [4, 5, 25]. The key aim of this section is to approximate the optimal cost functional $V^*(x_k)$, using a NN-based algorithm. Applying NN-based algorithm, the cost-functional $V(x_k)$ is approximated as $\hat{V}(x_k)$. The approximate cost functional $\hat{V}(x_k)$ is used to compute the approximate control inputs $\hat{u}_k$ and $\hat{v}_k$. To estimate $\hat{V}(x)$ using NN, the basis function $\sigma(x_k) = [\sigma_1(x_k) \ \sigma_2(x_k) \cdots \sigma_l(x_k)]^T$ and weight vector $\hat{w} = [\hat{w}_1, \hat{w}_2, \hat{w}_3 \cdots \hat{w}_l]^T$ are selected. The scalar l denotes the number of hidden layers in the NN. The selection of activation function depends on the following polynomial [112, 17]

$$\sum_{j=1}^{L/2} \left(\sum_{k=1}^{n} x_k \right)^{2j} \tag{5.18}$$

where L and n represent the order of approximation and the dimension of the system, respectively. Equation (5.19) corresponds to the activation function for a 2-dimensional system as

$$\sigma(x_k) = \{x_1^2, x_1 x_2, x_2^2, x_1^2 x_2^2, x_1^4 \cdots x_2^L\}. \tag{5.19}$$

The selected basis function $\sigma(x_k)$ is smooth and continuous, moreover it also holds the property $\sigma(0) = 0, \forall x_k = 0$. Applying the basis function $\sigma(x_k)$ and NN weight

$\hat{w}$, the estimated cost functional reduces to

$$\hat{V}(x_k) = \sum_{j=1}^{l} \hat{w}_j \sigma_l(x_k) \tag{5.20}$$

with a residual error (e_r)

$$\text{DT-GHJB}\left(\hat{V} = \sum_{j=1}^{l} \hat{w}_j \sigma_l(x_k), \hat{u}_k, \hat{v}_k\right) \triangleq e_r.$$

Applying the least square method [44], the unknown weight vector of NN is updated such that it minimizes the residual error e_r. The minimization of residual error e_r is done by projecting e_r on $\dfrac{de_r}{dw}$ i.e., $\langle \dfrac{de_r}{dw}, e \rangle = 0$ where $\langle a, b \rangle = \int_\Omega abdx$ is the Lebesgue integral. Due to the difficulty in this integration process, $\hat{w}$ is approximated using a mesh having ρ points on Ω from Riemann integration theory. The mesh point ρ is selected as $\rho \leq l$ with a mesh size Δx. Adopting Riemann approximation of integration, the $\langle \dfrac{de_r}{dw}, e \rangle = 0$ can be expressed as

$$X\hat{w} + Y. \tag{5.21}$$

This helps to derive the weight update law with least square error minimizing rule as

$$\hat{w} = -(X^T X)^{-1}(XY). \tag{5.22}$$

where X and Y are defined as

$$X = \begin{bmatrix} \{\nabla \hat{V}^T (f + g\hat{u} + M\hat{v} - x) + \frac{1}{2}(f + g\hat{u} \cdots \\ + M\hat{v} - x)^T \nabla^2 \hat{V}(f + g\hat{u} + M\hat{v} - x)\} \mid_{x=x_1} \\ \vdots \\ \{\nabla \hat{V}^T (f + g\hat{u} + M\hat{v} - x) + \frac{1}{2}(f + g\hat{u} \cdots \\ + M\hat{v} - x)^T \nabla^2 \hat{V}(f + g\hat{u} + M\hat{v} - x)\} \mid_{x=x_\rho} \end{bmatrix} \tag{5.23}$$

$$Y = \begin{bmatrix} d_{max}^2(x_k) + v_{max}^2(x_k) + x(k)^T Q x(k) \cdots \\ + \hat{u}(k)^T R_1 \hat{u}(k) + \hat{v}(k)^T R_2 \hat{v}(k) \mid_{x=x_1} \\ \vdots \\ d_{max}^2(x_k) + v_{max}^2(x_k) + x(k)^T Q x(k) \cdots \\ + \hat{u}(k)^T R_1 \hat{u}(k) + \hat{v}(k)^T R_2 \hat{v}(k) \mid_{x=x_\rho} \end{bmatrix} \tag{5.24}$$

Using estimated weight (5.22), the cost-functional is also estimated by applying the equation (5.20). The estimated cost-functional (5.20) is applied to derive the approximated control inputs $\hat{u}_k$ and $\hat{v}_k$. An algorithmic representation of numerical steps to achieve the suboptimal inputs $\hat{u}_k^*$ and $\hat{v}_k^*$ are given next.

Algorithm 3 Optimal inputs using NN-based approximation

1: Initialization: $k \Leftarrow 0$, $i \Leftarrow 0$, $x \Leftarrow x_0$, $u \Leftarrow u_0$, $v \Leftarrow v_0$.
2: Select any value of a scalar $\varepsilon > 0$ and number of mesh points ρ.
3: Initial inputs u_0 and v_0 are admissible control inputs.
4: Create an NN as $\hat{V}(x)$ using (5.20).
5: Compute $\hat{V}_i$ using (5.20), (5.22), (5.23), and (5.24). Here $i = 0, 1, 2, \ldots$, denotes the number of iteration.
6: Compute the approximate control inputs using following equation

$$
\begin{bmatrix} \hat{u}_{i+1} \\ \hat{v}_{i+1} \end{bmatrix} = - \left[\begin{array}{c|c} (2R_1 + g^T \nabla^2 \hat{V}_i g) & g^T \nabla^2 \hat{V}_i M \\ \hline M^T \nabla^2 \hat{V}_i g & (2R_2 + M^T \nabla^2 \hat{V}_i M) \end{array} \right]^{-1}
\begin{bmatrix} g^T (\nabla \hat{V}_i + \nabla^2 \hat{V}_i (f-x)) \\ M^T (\nabla \hat{V}_i + \nabla^2 \hat{V}_i (f-x)) \end{bmatrix}
\tag{5.25}
$$

7: Update the control inputs (5.25).
8: **if** $\hat{V}_i - \hat{V}_{i+1} \geq \varepsilon$ **then**
9:     Go to line 5
10: **else**
11:     Optimal cost-function $\hat{V}^* = \hat{V}_i$
12:     Using $\hat{V}^*$, the approximate optimal inputs $\hat{u}_k^*$ and $\hat{v}_k^*$ are computed as

$$
\begin{bmatrix} \hat{u}_k^* \\ \hat{v}_k^* \end{bmatrix} = - \left[\begin{array}{c|c} (2R_1 + g^T \nabla^2 \hat{V}^* g) & g^T \nabla^2 \hat{V}^* M \\ \hline M^T \nabla^2 \hat{V}^* g & (2R_2 + M^T \nabla^2 \hat{V}^* M) \end{array} \right]^{-1}
\begin{bmatrix} g^T (\nabla \hat{V}^* + \nabla^2 \hat{V}^* (f-x)) \\ M^T (\nabla \hat{V}^* + \nabla^2 \hat{V}^* (f-x)) \end{bmatrix}
\tag{5.26}
$$

13: **end if**

Remark 21 *Given admissible control inputs $u_0 \in \Omega_u$ and $v_0 \in \Omega_v$, the solution $\hat{V}_i$ of DT-GHJB (5.12) iteratively converges to its optimal solution V^* by updating the control inputs using (5.25). This claim can be proved analytically using the results reported in [112, 25].*

5.2.2 STABILITY OF UNCERTAIN SYSTEMS USING APPROXIMATE INPUTS

The derived approximate optimal inputs (5.26) for (5.8) ensure the asymptotic stability of uncertain system (5.3). This information is stated as a theorem in below.

Theorem 12 *Suppose there exists a continuously differentiable positive function $\hat{V}^*(x_k)$ which satisfies (5.12) with the inequality*

$$
d_{max}^2 \geq \phi^T \{ R_2 + (g^+ S)^T (R_1 + g^T \nabla^2 \hat{V}^* g)(g^+ S) + M^T \nabla^2 \hat{V}^* M \} \phi.
\tag{5.27}
$$

The approximated optimal control input $\hat{u}_k^*$ defined in (5.26) for (5.8) will be the robust solution of unmatched system (5.3) if the following condition holds

$$v_{max} \geq \hat{v}_k^{*T}(2R_2 + M^T\nabla^2\hat{V}^*M)\hat{v}_k^*. \tag{5.28}$$

Proof of Theorem 12 Let $V(x_k)$ is the solution of (5.12) and it is approximated as $\hat{V}^*(x_k)$ using the estimated inputs (5.26). The approximated solution $\hat{V}^*(x_k)$ and inputs (5.26) also satisfy the following equation

$$d_{max}^2(x_k) + v_{max}^2(x_k) + x_k^T Q x_k + \hat{u}_k^{*T}R_1\hat{u}_k^* + \hat{v}_k^{*T}R_2\hat{v}_k^* +$$
$$\nabla\hat{V}_k^{*T}(x_{k+1} - x_k) + \frac{1}{2}(x_{k+1} - x_k)^T\nabla^2\hat{V}_k^*(x_{k+1} - x_k) = 0. \tag{5.29}$$

Now, with the control inputs (5.26), the difference of $\hat{V}^*(x_k)$ $[\Delta\hat{V}^* = \hat{V}^*(x_{k+1})) - \hat{V}^*(x_k)]$ along the solution of (5.3) is

$$\Delta\hat{V}^* = \nabla\hat{V}^{*T}(f + g\hat{u}_k^* + M\hat{v}_k^* - x) + \frac{1}{2}(f + g\hat{u}_k^* + M\hat{v}_k^* - x)^T\nabla^2\hat{V}^*(f + g\hat{u}_k^* + M\hat{v}_k^* - x)$$
$$+ \nabla\hat{V}^{*T}(d - M\hat{v}_k^*) + (f + g\hat{u}_k^* + M\hat{v}_k^* - x)^T\nabla^2\hat{V}^*(d - M\hat{v}_k^*)$$
$$+ \frac{1}{2}(d - M\hat{v}_k^*)^T\nabla^2\hat{V}^*(d - M\hat{v}_k^*). \tag{5.30}$$

Using (5.7) in (5.30), the following is obtained

$$\Delta\hat{V}^* = \nabla\hat{V}^{*T}(f + g\hat{u}_k^* + M\hat{v}_k^* - x) + \frac{1}{2}(f + g\hat{u}_k^* + M\hat{v}_k^* - x)^T\nabla^2\hat{V}^*(f + g\hat{u}_k^* + M\hat{v}_k^* - x)$$
$$+ \nabla\hat{V}^{*T}(gN\phi + M\phi - Mv_k^*) + (f + g\hat{u}_k^* + M\hat{v}_k^* - x)\nabla^2\hat{V}^*(gN\phi + M\phi - M\hat{v}_k^*)$$
$$+ \frac{1}{2}(gN\phi + M\phi - M\hat{v}_k^*)^T\nabla^2\hat{V}^*(gN\phi + M\phi - M\hat{v}_k^*). \tag{5.31}$$

where matrix $N = g^+S$. After further simplification, equation (5.26) can be rewritten as

$$g^T\nabla^2\hat{V}^*(f + g\hat{u}_k^* + M\hat{v}_k^* - x) = -(2R_1\hat{u}_k^* + g^T\nabla\hat{V}^*), \tag{5.32}$$
$$M^T\nabla^2\hat{V}^*(f + g\hat{u}_k^* + M\hat{v}_k^* - x) = -(2R_2\hat{v}_k^* + M^T\nabla\hat{V}^*). \tag{5.33}$$

Applying (5.29), (5.32), and (5.33) in (5.31), $\Delta\hat{V}^*$ is simplified as

$$\Delta\hat{V}^* = -d_{max}^2 - v_{max}^2 - x_k^T Q x_k - \hat{u}_k^{*T}R_1\hat{u}_k^* - \hat{v}_k^{*T}R_2\hat{v}_k^* + \nabla\hat{V}^{*T}(gN\phi + M\phi - M\hat{v}^*)$$
$$- (2R_1\hat{u}^* + g^T\nabla\hat{V}^*)^Tg^+S\phi - (2\hat{v}^{*T}R_2 + \nabla^T\hat{V}^*M)(\phi - \hat{v}^*)$$
$$+ \frac{1}{2}(gN\phi + M\phi - M\hat{v}^*)^T\nabla^2\hat{V}^*(gN\phi + M\phi - M\hat{v}^*).$$

After further simplification $\Delta\hat{V}^*$ reduces to

$$\Delta\hat{V}^* \leq -x_k^T Q x_k - \{v_{max}^2 - \hat{v}_k^{*T}(2R_2 + M^T\nabla^2\hat{V}^*M)\hat{v}_k^*\}$$
$$- \{d_{max}^2 - \phi^T(R_2 + N^T(R_1 + g^T\nabla^2\hat{V}^*g)N + M^T\nabla^2\hat{V}^*M)\phi\}. \tag{5.34}$$

From the above inequality, $\Delta\hat{V}^*$ is negative definite if the conditions (5.27) and (5.28) hold. This proves the asymptotic convergence of (5.3) under periodic feedback of control input $u_k, \forall k.$

Remark 22 *It is observed that the DT-HJB (5.10) is approximated using Taylor series expansion and it reduces to DT-GHJB (5.12). Due to this approximation, the optimal input (5.11) is converted to suboptimal input (5.16). The approximated virtual input $\hat{v}_k^*$ is not used to stabilize system (5.3) but it is used to design actual control input $\hat{u}_k^*$. The input $\hat{v}_k^*$ is used to verify the condition (5.28).*

The proposed robust control framework considers the general system uncertainty, which include both matched and mismatched component. Without mismatched part, system (5.3) reduces to matched system (defined in (5.5)), i.e.,

$$x_{k+1} = f(x_k) + g(x_k)(u_k + d(x_k)). \tag{5.35}$$

Moreover, due to the absence of mismatched part, the virtual control input v_k is not necessary in (5.8) and (5.9). Therefore the nominal system and cost-functional for (5.35) reduce to

$$x_{k+1} = f(x_k) + g(x_k)u_k(x_k) \tag{5.36}$$

$$J_k = \frac{1}{2}\sum_{k=0}^{\infty}\left\{d_{max}^2(x_k) + x_k^T Q x_k + u_k^T R_1 u_k\right\} \tag{5.37}$$

where $\|d(x_k)\| \le d_m(x_k) \; \forall k$. As a special case of Theorem 12, Corollary 2 is introduced for matched system.

Corollary 2 *Suppose there exists a continuously differentiable positive function $\hat{V}^*(x_k)$ which satisfies*

$$d_{max}^2(x_k) + x_k^T Q x_k + \hat{u}_k^{*T} R_1 \hat{u}_k^* + \nabla\hat{V}^{*T}(x_{k+1} - x_k)$$
$$+ \frac{1}{2}(x_{k+1} - x_k)^T \nabla^2\hat{V}^*(x_{k+1} - x_k) = 0 \tag{5.38}$$

Then the designed optimal control input

$$\hat{u}_k^* = -\{g(x_k)^T \nabla^2\hat{V}^* g(x_k) + 2R_1\}^{-1} g(x_k)^T \{\nabla\hat{V}^* + \nabla^2\hat{V}^*(f - x_k)\} \tag{5.39}$$

for (5.36) which minimizes (5.37) is also a robust solution of (5.35) if the uncertainty $d(x_k)$ satisfies the following bound

$$d_{max}^2(x_k) \ge \phi^T(N^T(R_1 + g^T\nabla^2\hat{V}^* g)N)\phi. \tag{5.40}$$

Proof *The proof of this corollary is straightforward and hence it is omitted here.*

5.2.3　WITH INPUT UNCERTAINTY

The proposed framework can be extended in the presence of input uncertainty. A system with mismatched input uncertainty is described as

$$x_{k+1} = f(x_k) + \{g(x_k) + d(x_k)\}u_k(x_k) \tag{5.41}$$

where function $d(x_k)$ is the bounded uncertainty affecting the input function $g(x_k)$. To design the robust control input the virtual nominal system (5.8) and cost-functional (5.9) are considered. To tackle the mismatched uncertainty in input function, the optimal control problem is solved for (5.8) and (5.9) with the control inputs (5.26). The results of robust problem for an input uncertain system is presented through the following theorem.

Theorem 13 *Suppose there exists a continuously differentiable positive function $\hat{V}^*(x_k)$ which satisfies (5.29) with the inequality*

$$\begin{aligned} d_{max}^2 \quad \geq \quad & (\phi u_k^*)^T \{R_2 + (g^+ S)^T R_1 (g^+ S) + M^T \nabla^2 \hat{V}^* M \\ & + (gg^+ S + M)^T \nabla^2 \hat{V}^* (gg^+ S + M)\}(\phi u_k^*). \end{aligned} \tag{5.42}$$

The approximate optimal control input $\hat{u}_k^*$ defined in (5.26) for (5.8) which minimizes (5.9) will be the robust solution of (5.41) if the following condition holds

$$v_{max} \geq \hat{v}_k^{*T}(2R_2 + M^T \nabla^2 \hat{V}^* M)\hat{v}_k^*. \tag{5.43}$$

Proof *The proof of the Theorem 13 is similar to the proof of Theorem 12, hence it is omitted here.*

5.2.4　COMPARISON WITH EXISTING RESULTS

This subsection compares the main results of this chapter with the existing work reported in [132]. In 2016, D. Wang et al. have proposed an approximate optimal control based robust control technique for discrete-time nonlinear system. To realize the robust control law, they have considered that the system is affected by matched uncertainty. For the purpose of comparison with the results described in [132], the mismatched component of the uncertainty is neglected. It is observed that without mismatched component, the virtual input v_k is not necessary. Therefore without virtual input v_k, the nominal dynamics and cost-functional defined in this chapter are in a form similar to that as mentioned in [132].

So the results reported in [132] can be shown to be a special case of the proposed work. To solve the nonlinear optimal control problem, a NN-based approximation technique is adopted from [132], [25]. But, the presence of control input v_k in nominal system (5.8) modifies the DT-HJB equation reported in [132], [25]. To tackle the mismatched uncertainty, the cost-functional (5.9) consists of two extra terms as $v_{max}^2(x_k)$ and $v_k^T R_2 v_k$. These two extra terms directly affect the computation of matrices X and Y as mentioned in (5.23) and (5.24) respectively. Moreover, the computation of approximated cost $\hat{V}^*$ also depends on both the control inputs $\hat{u}_k^*$ and $\hat{v}_k^*$. The absence of virtual input v_k in [132], [25], makes it easier to compute $\hat{V}^*$ over the proposed approach as reported in Algorithm 3.

5.3   SIMULATION RESULTS

The section uses a numerical example to validate the proposed control algorithm. Consider a state space form of uncertain discrete-time nonlinear system as (5.3) where functions $f(x_k)$, $g(x_k)$, and $\phi(x_k)$ are defined as

$$f(x_k) = \begin{bmatrix} -0.8x_{2k} \\ sin(0.8x_{1k}) - x_{2k} + 1.8x_{2k} \end{bmatrix}, \quad g(x_k) = \begin{bmatrix} 0 \\ -x_{2k} \end{bmatrix}$$
$$\text{and } \phi(x_k) = p \; sin(0.8k)x_{1k}.$$

Here p is the uncertain parameter. This system has mismatched uncertainty and hence the results of [132] are not applicable. To solve the optimal control problem for virtual nominal system, the design parameters $Q = I$, $R1 = 0.5I$ and $R2 = 0.5I$ are selected. The scaling matrix S is selected as $S = \begin{bmatrix} 0.1 & 0.2 \end{bmatrix}^T$. The upper bound of uncertainty $d(x_k)$, defined in (5.4) is considered as $d_{max} = \|x_k\|^2$. The parameter p can vary within -0.5 to 0.5. To estimate the optimal cost function through the NN realization, the NN is constructed as

$$\hat{V}(x) \quad = \quad \hat{w}_1 x_1^2 + \hat{w}_2 x_2^2 + \hat{w}_3 x_1 x_2 \tag{5.44}$$

The mesh point $\rho = 6$ and mesh size $\Delta x = 0.01$ are selected. For simulation, the initial admissible control inputs $u_0 = x_1 + 1.5x_2$ and $v_0 = 0.049x_1$ are used. The simulation is carried out in Matlab for 10 iterations with the initial states $[0.5, -0.5]^T$. After 5 iterations, the NN weight w converges to $w = \begin{bmatrix} 6.97 & 8.35 & 6.72 \end{bmatrix}^T$.

Analysis of simulation results

Figure 5.2(a) shows that the system has converged to its equilibrium point through the admissible control inputs u_0. Figures 5.3(a) and 5.3(b) show the convergence of NN weight and approximated value function. In Fig. 5.2(b), the systems state trajectories reach their equilibrium point in-spite-of uncertainty. The simulation results show that the proposed robust suboptimal control technique ensures the closed-loop stability in presence of mismatched uncertainty. The variation of stabilizing input $\hat{u}_k^*$ and virtual input $\hat{v}_k^*$ is shown in Fig. 5.4(a) and 5.4(b). Now, for a selection of the scaling matrix $S = g(x_k)$, the same example is solved numerically. This selection converts the mismatched system (5.3) to a matched system as defined in (5.35). The closed loop behavior of (5.35), is shown in Figs. 5.5(a)–5.5(b), which replicates the results of matched system as stated in [132].

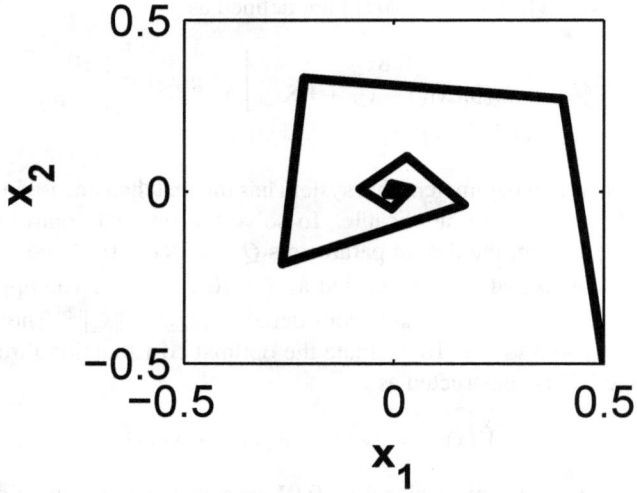

(a) Convergence of nominal system's states x_1 and x_2 with the initial admissible control inputs u_0 and v_0.

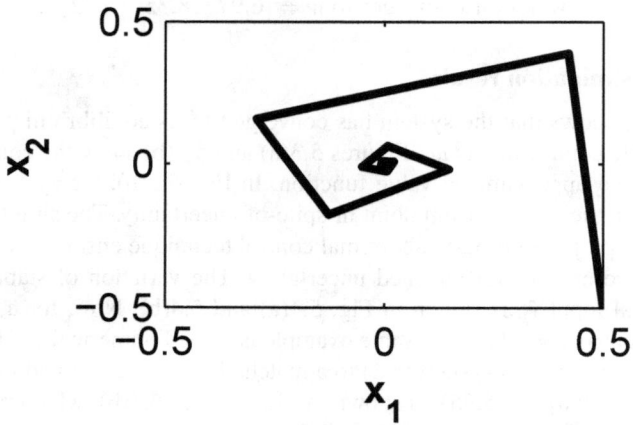

(b) Convergence of uncertain system's states x_1 and x_2 with the designed robust control input $\hat{u}_k^*$ for $p = 0.5$.

Figure 5.2 Results of the proposed robust control technique

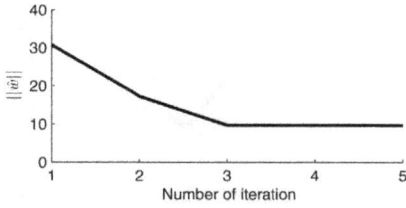

(a) Convergence of norm of weight vector ($\|\hat{w}\|$).

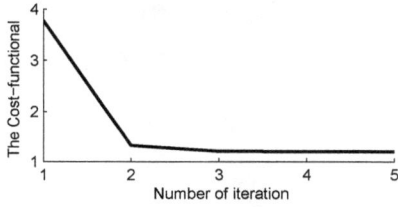

(b) Convergence of approximated cost-functional.

Figure 5.3 Results of NN-based approximation

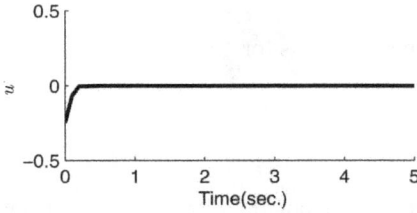

(a) Convergence of approximated control input $\hat{u}_k^*$.

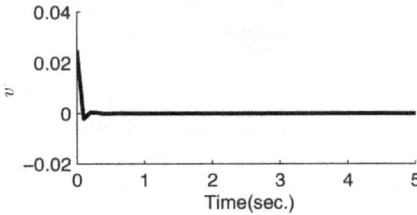

(b) Convergence of approximated virtual input $\hat{v}_k^*$.

Figure 5.4 Convergences of control inputs

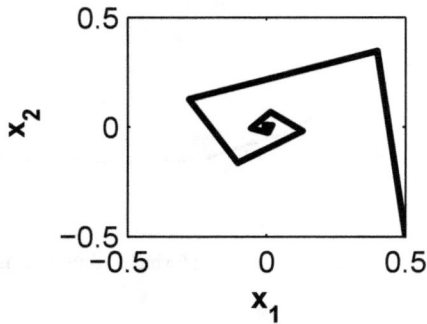

(a) Convergence of system states with matched uncertainty for $p = 0.5$.

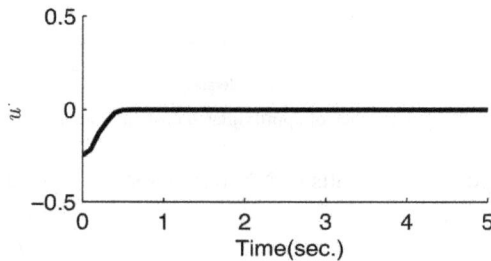

(b) Convergence of approximated control input $\hat{u}_k^*$ for matched system.

Figure 5.5 Results for matched uncertain system

5.4 SUMMARY

A discrete-time robust control technique for an uncertain nonlinear system is proposed in this chapter. It is considered that the system is affected by mismatched uncertainty. The control law is designed by formulating an optimal control problem for a virtual system with a quadratic cost-functional. The virtual system has two control inputs. Here u is stabilizing and v is virtual inputs. The virtual input v is defined to design the stabilizing control law u along with the stability condition. An analytical proof for ensuring asymptotic convergence of closed-loop uncertain system is also given. A comparative study between the existing and proposed results is also reported. It is shown that the existing results are the special case of the proposed work. In this chapter, the robust control law is derived with the periodic feedback of state information. It will be interesting to modify the proposed robust control framework with the event-based feedback of system state. In next chapter, the results of proposed robust control algorithm with event-triggered feedback are used to regulate the closed-loop behavior of different class of systems.

Section III

Applications

6 Applications

After studying this chapter, one should be able to: design a robust controller with matched and mismatched uncertainties; design a robust event-triggered controller for various nonlinear systems, mainly robotic systems, Lipschitz systems and batch reactor systems; establish the stability results through ISS theory for a class nonlinear systems under presence of system uncertainties and intermittent feedback; validate the control approach numerically and show the effectiveness of proposed event-triggered robust control laws.

6.1 INTRODUCTION

In this chapter, the proposed control algorithms discussed in Chapter 2 are used to stabilize different class of systems with limited feedback information. Section 6.2 considers a robot manipulator dynamics (Euler-Lagrange system) for design and analysis. In Section 6.3, the proposed idea is applied on a class of Lipschitz system. These systems are inherently nonlinear. Attempt is made to rewrite the system dynamics as a linear model with uncertainty. In these formulations, the system non-linearities and parametric variation of system model are considered as a source of uncertainties. An event-based linear robust control algorithm is proposed in order to stabilize these class of nonlinear systems with limited feedback information. Moreover, an event-based control algorithm is proposed in Section 6.4 to regulate the state of a batch reactor system. It is considered that the reactor system is affected by norm bounded mismatched uncertainty. To stabilize such a system, a robust control law is derived based on the nominal dynamics and the prior knowledge of uncertainty bound. Next, the derived controller gain matrix is used to analyze the closed loop performance. To illustrate the proposed control algorithm, the simulation results are reported. The primary contributions of this chapter are listed below:

Chapter 2 proposed an event-triggered control algorithm for an uncertain linear system. In this chapter, the Euler-Lagrange and Lipschitz nonlinear systems are considered for analysis. The dynamics of these systems are divided into two parts linear and nonlinear one. In the manipulator example, the nonlinear part is treated as a source of matched uncertainty which directly affects the linear component of system dynamics. In case of Lipschitz nonlinear system, system nonlinearity is expressed as mismatched uncertainty. Using the optimal control framework for robust controller design, a linear control law is derived by solving a LQR problem. To tackle such uncertainties, the Riccati equation and cost-functional are redesigned

DOI: 10.1201/9781003229698-6

and compared with existing work reported in Chapter 2. The linear robust control law ensures the ISS of original nonlinear system.

In Chapter 2, a control algorithm has been proposed for a mismatched linear system. For the purpose of analysis, the system uncertainty has subdivided into a matched and mismatched component as mentioned in (2.4). In this chapter [Section 6.4], the norm bounded mismatched uncertainty is expressed by three matrices and out of them two are known. This uncertainty description helps to simplify the results.

6.2   ROBUST EVENT-TRIGGERED CONTROL OF ROBOT MANIPULATOR

A two-link SCARA type robot manipulator is shown in Fig. 6.1. For simplicity, it

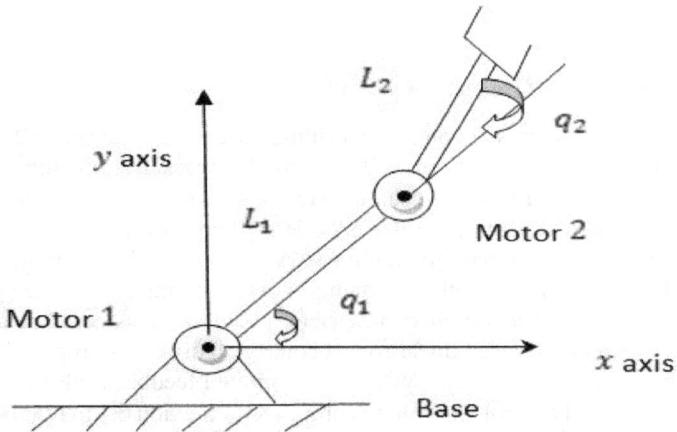

Figure 6.1 Two-link robot manipulator

is assumed that the link masses are concentrated at the center of masses of the arm. The system parameters are defined next.

From [29] and [3], the dynamics of robot manipulator is written as

$$M(q)\ddot{q}+N(q,\dot{q}) = \tau \tag{6.1}$$

where $N(q,\dot{q}) = V(q,\dot{q})+F(\dot{q})+G(q)$. The vectors $q(=[q_1,q_2]^T)$ and $\tau(=[\tau_1,\tau_2]^T)$ denote the joint variables and generalized forces, respectively. The inertia matrix, Coriolis vector, gravity vector, and friction vector of (6.1) are represented as $M(q)$, $V(q,\dot{q}), G(q)$ and $F(\dot{q})$ respectively. Due to unknown load and unmodeled frictions, uncertainty may be introduced in $M(q)$ and $N(q,\dot{q})$ of (6.1). The state space representation of (6.1) with system uncertainty is

$$\dot{x} = \begin{bmatrix} \dot{x}_1 \\ \dot{x}_2 \end{bmatrix} = \begin{bmatrix} \dot{q}_1 \\ \dot{q}_2 \end{bmatrix} = Ax + B(u+h(x)u) + Bf(x) \tag{6.2}$$

Table 6.1
Robot Parameters

Parameter	Name
L_1	length of Link 1
L_2	length of Link 2
q_1	joint angle of link 1
q_2	joint angle of link 2
m_1	mass of link 1
m_2	mass of link 2
m_l	mass of load
τ_1	torque applied to joint 1
τ_1	torque applied to joint 2

where $A = \begin{bmatrix} 0 & I \\ 0 & 0 \end{bmatrix}$ and $B = \begin{bmatrix} I \\ 0 \end{bmatrix}$. The two nonlinear functions $h(x)$ and $f(x)$ are treated as uncertainty sources. The input uncertainty $h(x)$ and system uncertainty $f(x)$ are defined as

$$h(x) = M(x_1)^{-1} M_0(x_1) - I \geq 0, \tag{6.3}$$

$$f(x) = M(x_1)^{-1}(N_0(x_1, x_2) - N(x_1, x_2)). \tag{6.4}$$

It is assumed that the function $f(x)$ satisfies the following assumption:

Assumption 3 *For a positive definite matrix F_m, the uncertainty $f(x)$ satisfies*

$$f(x)^T f(x) \leq x^T F_m x. \tag{6.5}$$

To achieve the robust control solution $u(x)$ of (6.2), we need to solve an optimal control problem as per *Lemma 3*. The nominal system (2.7) with the modified cost-functional (2.8) are considered to design the optimal feedback control law $u(x)$. **From now onwards, the controller gain K_1 and control input u_1 [mentioned in (2.7)] are denoted as K and u, respectively. For simplicity, the matrix $R = I$ is selected.** Based on optimal control theory, the optimal input $u(x)$ satisfies the equation (2.20) and (2.22).

To reduce the total number of actuation, an event-based control input is applied, which reduces the robot dynamics as

$$\dot{x} = Ax + B\{u(t_k) + h(x)u(t_k)\} + Bf(x). \tag{6.6}$$

The event-triggered control input replaces the continuous state-feedback control law $u(t)(= Kx(t))$ by

$$u(t_k) = Kx(t_k). \tag{6.7}$$

Here the sequence t_k represents aperiodic state transmission, control input computation and actuation instant. From [117], the event-based closed loop system can be written as

$$\dot{x} = Ax + B\{(Kx + Ke) + h(x)(Kx + Ke)\} + Bf(x) \tag{6.8}$$

where variable $e(t)$ is the measurement error and it is defined in (2.16). Now the primary aim is to show that the system (6.8) is ISS with-respect to measurement error e. To achieve this aim, a robust control law and an event-triggering sequence are proposed which ensure the ISS of (6.8). The main results are stated in the form of following theorem:

Theorem 14 *Suppose there exist two scalars $\sigma \in (0,1)$ and $\varepsilon > 0$ and the controller gain K is designed for the nominal system (2.7) by minimizing the cost-functional (2.8). The uncertain system (6.8), with event-triggered control law (6.7), is ISS if there exists an event occurring sequence $\{t_k\}_{k \in I}$ given by*

$$t_0 = 0, \ t_{k+1} = inf\{t \in \mathbb{R} | t > t_k \wedge \mu\|x\|^2 - \|e\|^2 \leq 0\} \tag{6.9}$$

where variable $\mu \left(= \frac{\sigma \lambda_{min}(Q_1)}{\|K^T K + K^T h^T hK\|} \right)$ is a design parameter and matrix $Q_1 = (Q - 2K^T K) > 0$.

Proof of Theorem 14 *To ensure asymptotic stability of the closed loop system (6.8), it is necessary to simplify $\dot{V}(x)$ in the form of (1.35). Suppose $V(x) = x^T Px$ is the Lyapunov function for (6.8). To prove the asymptotic stability of (6.8), the time derivative of $V(x)$ can be simplified as*

$$\dot{V}(x) = V_x^T (Ax + B\{(Kx + Ke) + h(x)(Kx + Ke)\} + Bf(x)). \tag{6.10}$$

Using (2.20) and (2.22), equation (6.10) is simplified as

$$\begin{aligned} \dot{V}(x) &= -(x^T F_m x + x^T Qx + 2x^T K^T hKx + 2x^T (K^T K + K^T hK)e \\ &\quad + (u^* + f(x))^T (u^* + f(x))) + f(x)^T f(x) \\ &\leq -\lambda_{min}(Q_1)\|x\|^2 + \|K^T K + K^T h^T hK\|\|e\|^2 \end{aligned} \tag{6.11}$$

where matrix $Q_1 = Q - K^T K > 0$. From (6.11) and (1.35), it can be written that the control input updation is required only when the following triggering condition is violated:

$$\|e\|^2 \leq \mu\|x\|^2 \tag{6.12}$$

where parameter μ is

$$\mu = \frac{\sigma \lambda_{min}(Q_1)}{\|K^T K + K^T h^T hK\|}, \ \forall \sigma \in (0,1). \tag{6.13}$$

Using (6.12) and (6.13), the event-triggering instants are defined in (6.9). The event-triggering sequence (6.9) ensures the ISS of (6.8) with the limited feedback information.

6.2.1 SIMULATION RESULTS

This subsection illustrates the above introduced theoretical results with an example of two-link SCARA type robot manipulator. For simulation, the state and system matrices of two-link manipulator are $x = [x_1 \ x_2 \ x_3 \ x_4]^T = [q_1 \ q_2 \ \dot{q}_1 \ \dot{q}_2]^T$,

$$A = \begin{bmatrix} 0 & 0 & 1 & 0 \\ 0 & 0 & 0 & 1 \\ 0 & 0 & 0 & 0 \\ 0 & 0 & 0 & 0 \end{bmatrix} \text{ and } B = \begin{bmatrix} 0 & 0 \\ 0 & 0 \\ 1 & 0 \\ 0 & 1 \end{bmatrix}. \text{ Here } q_1, q_2, \text{ represent the joint angle of first}$$

and second link and $\dot{q}_1$, $\dot{q}_2$ represent corresponding angular velocity. Based on (6.4) and (6.3), the expression of system and input uncertainties $f(x)$ and $h(x)$ are given

$$f(x) = \begin{bmatrix} 562.0 + 171.6\cos x_2 + 100m_L + 96m_L\cos x_2 & 51.2 + 85.8\cos x_2 + 36m_L + 48m_L\cos x_2 \\ 51.2 + 85.8\cos x_2 + 36m_L + 48m_L\cos x_2 & 51.2 + 36m_L \end{bmatrix}^{-1}$$
$$\times \begin{bmatrix} 48m_L(2x_3 - x_4)x_4\sin x_2 + 78.4m_L\sin x_1 + 58.8m_L\sin(x_1 + x_2) \\ 48m_Lx_3{}^2\sin x_2 + 58.8m_L\sin(x_1 + x_2) \end{bmatrix} \quad (6.14)$$

$$h(x) = \begin{bmatrix} 562.0 + 171.6\cos x_2 + 100m_L + 96m_L\cos x_2 & 51.2 + 85.8\cos x_2 + 36m_L + 48m_L\cos x_2 \\ 51.2 + 85.8\cos x_2 + 36m_L + 48m_L\cos x_2 & 51.2 + 36m_L \end{bmatrix}^{-1}$$
$$\begin{bmatrix} 2562.0 + 2091.6\cos x_2 & 771.2 + 1045.8\cos x_2 \\ 771.2 + 1045.8\cos x_2 & 771.2 \end{bmatrix} - I \quad (6.15)$$

in (6.14) and (6.15), where m_L is used to denote the mass of unknown load. The primary objective of this control technique is to calculate an optimal gain K for (2.7), which minimizes (2.8). Here matrices F_m and Q are considered as

$$F_m = \begin{bmatrix} 27.6 & 14.3 & 198.6 & -81.7 \\ 14.3 & 8.6 & 105.3 & -35.1 \\ 198.6 & 105.3 & 1432.9 & -573.2 \\ -81.7 & -35.1 & -573.2 & 286.6 \end{bmatrix} \text{ and } Q = (2 \times 10^3)I.$$

Due to linear nominal system, the optimal control problem turns to an infinite-time LQR problem, which motivates us to solve an ARE, $A^TP + PA + F + 10I - PBB^TP = 0$. The positive definite solution of this Riccati equation P is used to calculate the optimal control input $u = \begin{bmatrix} -4.98 & -1.92 & -36.14 & 11.71 \\ -1.92 & -2.42 & 11.71 & -12.46 \end{bmatrix} x$. To realize the event-triggering sequence (6.9), the design parameter $\sigma = 0.6$ is selected. The simulation is executed for 25 seconds with initial state vector $(\pi/4, \pi/2, 0, 0)$. Figures 6.2(a) – 6.2(d), show the convergence of system states for different values of m_L. Unknown load m_L can be selected any value between 10 oz and 20 oz. On the other hand, the Fig. 6.3 shows the control input and its updating instant for $m_L = 20$ oz. Fig. 6.4 shows the zoomed view of Figure 6.3 for first few seconds. It is seen from Fig. 6.4 that number of computation is significantly reduced which also ensures the reduction of transmission cost over the communication channel.

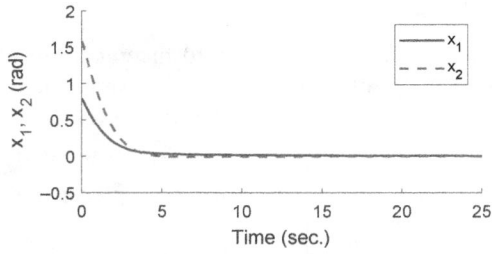

(a) Angle positions of two-link manipulator for $m_L = 10$ oz.

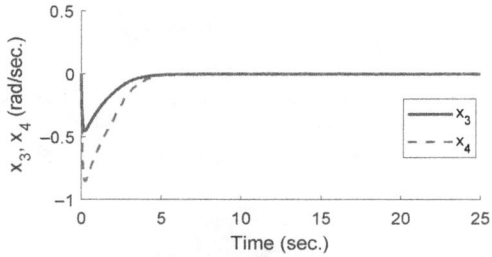

(b) Angle velocities of two-link manipulator for $m_L = 10$ oz.

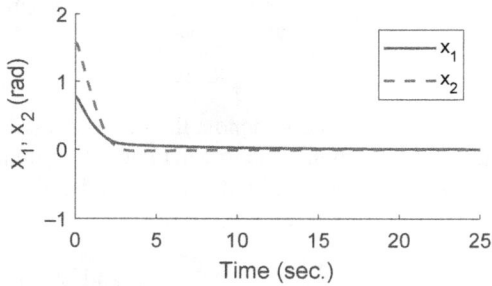

(c) Angle positions of two-link manipulator for $m_L = 20$ oz.

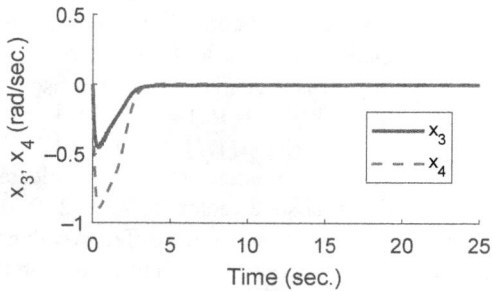

(d) Angle velocities of two-link manipulator for $m_L = 20$ oz.

Figure 6.2 Stabilization of states for different value of m_L

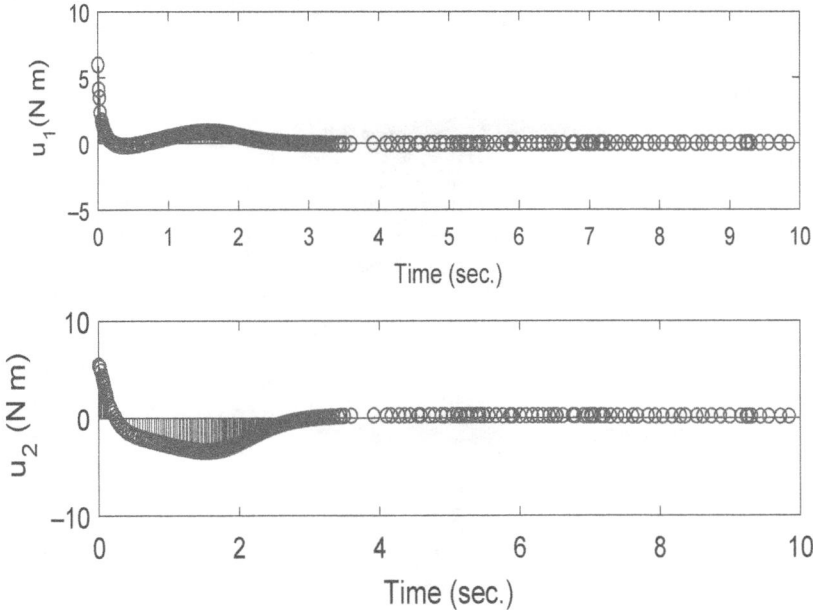

Figure 6.3 Control input at respective event-generating instant for $m_L = 20$ oz

6.3 ROBUST EVENT-TRIGGERED CONTROL OF LIPSCHITZ NONLINEAR SYSTEMS

This section describes an event-triggered control algorithm for a Lipschitz nonlinear system. Consider an event-triggered Lipschitz nonlinear system as

$$\dot{x}(t) = Ax(t) + Bu(t_k) + D\Phi(x) \qquad (6.16)$$
$$u(t_k) = K(x + e) \qquad (6.17)$$

where $x \in \mathbb{R}^n$, $u \in \mathbb{R}^m$ are system state and input vectors, respectively. The matrix K represents the controller gain which is derived based on the optimal control theory. The error variable e is defined in (2.16). The nonlinear function $\Phi(x)$ holds the following property

$$\|\Phi(X_1) - \Phi_2(X_2)\| \le L_c \|X_1 - X_2\|, \ \forall \ (X_1, X_2) \in \mathbb{R}^n \times \mathbb{R}^n. \qquad (6.18)$$

where scalar L_c represents the Lipschitz constant for the function $\Phi(x)$. The nonlinear function $\Phi(x)$ is treated as the source of uncertainty. The system (6.16) is a mismatched system as matrix $D \ne B$. The objective of this problem is stated below:

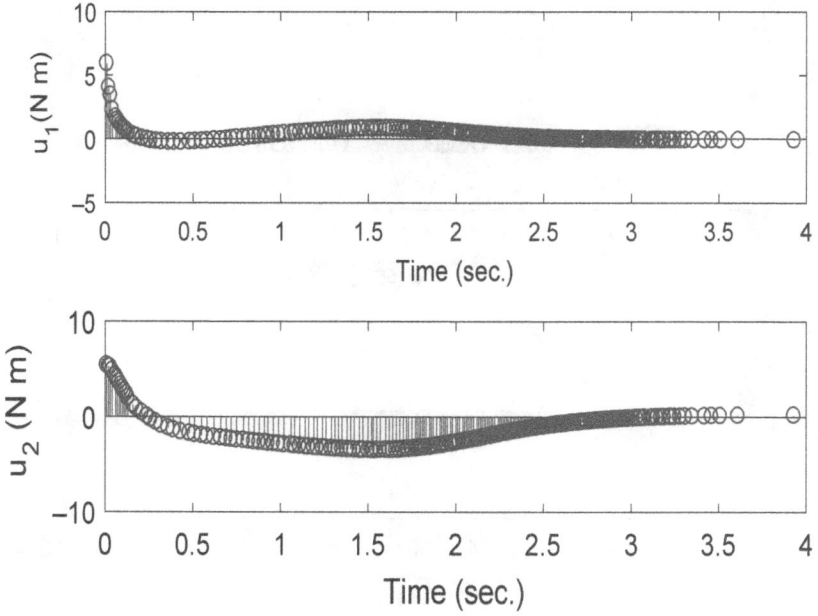

Figure 6.4 Zoomed view of control input at respective event-generating instant for $m_L = 20$ oz

Control objective: Design an event-triggered state feedback control law (6.17) which can ensure the ISS of closed loop system (6.16) in spite of system nonlinearity $D\Phi(x)$.

To derive the controller gain K, the steps mentioned in Chapter 2 are adopted. Using Lemma 3, the nominal system and cost-functional are selected as

$$\dot{x} = Ax + Bu + (I - BB^+)Dv \tag{6.19}$$

$$J = \int_0^\infty \frac{1}{2}(x^T(F + \eta^2 I)x + u^T u + \rho^2 v^T v)dt \tag{6.20}$$

where the scalar $x^T F x$ is used to define an upper bound such that

$$\Phi(x)^T[D^T B^{+T} B^+ D + I]\Phi(x) \leq x^T F x. \tag{6.21}$$

The notation B^+ is used to represent the pseudo inverse of input matrix B. Based on optimal control theory, the optimal inputs u and v should satisfy the following Riccati equation

$$PA + A^T P - PBB^T P - \rho^2 P(I - BB^+)DD^T(I - BB^+)^T P + F + \eta^2 I = 0. \tag{6.22}$$

The solution $P > 0$ of (6.22) is used to compute the optimal control inputs u and v as

$$u = \underbrace{-B^T P x}_{K} = Kx, \tag{6.23}$$

$$v = \underbrace{-\rho^{-2} D^T (I - BB^+)^T P x}_{L} = Lx \tag{6.24}$$

where K and L are gain matrices. The aperiodic sate information $x(t_k)$ and controller gain matrix K are used to derive the event-triggered control law (6.17). In order to derive the triggering condition and stability results, following theorem is introduced:

Theorem 15 *Let there exist scalars σ, η, β, and a positive-definite solution $P > 0$ of Riccati equation (6.22). Using the matrix P, the optimal gain matrices K and L are derived based on (6.23) and (6.24), respectively. The event-triggered control input (6.17) ensures the ISS of closed loop system (6.16) if the control input (6.17) is actuated based on the event-triggering sequence*

$$t_0 = 0, \ t_{k+1} = inf\{t \in \mathbb{R} | t > t_k \wedge \mu \|x\|^2 - \|e\|^2 \leq 0\} \tag{6.25}$$

with the following condition

$$2\rho^2 L^T L \leq \beta^2 I < \eta^2 I. \tag{6.26}$$

The scalar μ is defined as

$$\mu = \frac{\sigma(\eta^2 - \beta^2)^2}{4\|K^T K\|^2}, \ \forall \sigma \in (0,1). \tag{6.27}$$

Proof of Theorem 15 *Let $V(x) = x^T P x$ is a Lyapunov function for (6.16). Then, $\dot{V}(x)$ along the direction of (6.16) is*

$$\begin{aligned} \dot{V}(x) &= V_x^T (Ax + Bu + (I - BB^+)Dv) - V_x^T (I - BB^+)Dv + V_x^T D\Phi(x) + V_x^T BKe. \\ &= x^T (PA + A^T P - PBB^T P - \rho^2 P(I - BB^+)DD^T (I - BB^+)^T P)x \\ &\quad - x^T P(I - BB^+)Dv + 2x^T PD\Phi(x) + 2x^T PBKe \end{aligned} \tag{6.28}$$

where matrix V_x is used to represent $\frac{\partial V}{\partial x}(= 2Px)$. Using equations (6.22), (6.23), and (6.24), (6.28) is simplified as

$$\begin{aligned} \dot{V}(x) &= -\{x^T F x + \eta^2 x^T x + u^T u + \rho^2 v^T v\} + 2\rho^2 v^T v + 2x^T PBB^+ D\Phi(x) \\ &\quad + 2x^T P(I - BB^+)D\Phi(x) - 2x^T K^T Ke. \end{aligned} \tag{6.29}$$

After further simplification following is obtained

$$\begin{aligned} \dot{V}(x) &= -\{x^T F x + \eta^2 x^T x + u^T u + \rho^2 v^T v\} + 2\rho^2 v^T v \\ &\quad -2u^T B^+ D\Phi(x) - 2\rho^2 v^T \Phi(x) - 2x^T K^T Ke, \\ &\leq -[x^T F x - \Phi(x)^T (D^T B^{+T} B^+ D + I)\Phi(x)] - x^T (\eta^2 I - 2\rho^2 L^T L)x - 2x^T K^T Ke. \end{aligned} \tag{6.30}$$

Now, using (6.21) and (6.26), (6.30) reduces to

$$
\begin{aligned}
\dot{V}(x) &\leq -(\eta^2 - \beta^2)\|x\|^2 - 2x^T K^T K e \\
&\leq -\frac{q}{2}\|x\|^2 + \frac{2}{q}\|K^T K\|^2\|e\|^2
\end{aligned}
\tag{6.31}
$$

where $q(= \eta^2 - \beta^2) > 0$. Based on the Definition 8, the inequality (6.31) ensures the ISS of (6.16). The event-triggering condition (6.25) is also derived from (6.31).

The theoretical results are illustrated next through a numerical example.

6.3.1 SIMULATION RESULTS

Consider the state space form of a one-link robot manipulator with revolute joints [101] is defined as (6.16) where matrices $A = \begin{bmatrix} 0 & 1 & 0 & 0 \\ -48.9 & -1.25 & 48.6 & 0 \\ 0 & 0 & 0 & 1 \\ 19.5 & 0 & -19.5 & 0 \end{bmatrix}$,

$B = \begin{bmatrix} 0 & 21.6 & 0 & 0 \end{bmatrix}^T$, $D = I$, and uncertainty $\Phi(x) = \begin{bmatrix} 0 & 0 & 0 & \gamma \sin(x_3) \end{bmatrix}^T$. The uncertainty $\Phi(x)$ holds the property (6.18). For the purpose of simulation, the scalar γ is selected as 0.33. The simulation is carried out in Matlab for 30 seconds. The initial value of state vector is $\begin{bmatrix} 0.1 & 0.01 & 0.2 & 0.3 \end{bmatrix}^T$. The controller gain matrices K and L are computed as $K = \begin{bmatrix} -0.7789 & -1.5464 & -0.1717 & -0.4710 \end{bmatrix}$ and

$$
L = \begin{bmatrix} 4.4810 & 0.0361 & -2.8618 & 0.7824 \\ 0 & 0 & 0 & 0 \\ -2.8618 & 0.0080 & 4.0432 & -0.5539 \\ 0.7824 & 0.0218 & -0.5539 & 0.5015 \end{bmatrix}, \text{ respectively. The design matrices}
$$

$F = 10I$ and $R = 4I$ are selected. Here I is an identity matrix with appropriate dimension. The design parameters $\eta = 2.2$, $\beta = 2$, and $\rho = 0.1$ are considered which will satisfy the condition (6.26). To realize the event-triggering law (6.25), the scalar $\mu = 0.018$ is computed based on (6.27). Figure 6.5(a) shows the convergence of state trajectories with event-triggered control input. A comparative study between continuous and event-triggered based control technique is shown in Table 6.2. It shows the efficacy of the proposed event-triggering technique over the continuous one in terms of total number of actuation for the given run time.

Table 6.2

Comparative Results of Event-triggered and Continuous Control

Control mechanism	$\tau_{max}(sec.)$	$\tau_{min}(sec.)$	u_{total}
Continuous control	0.05	0.05	600
Event-triggered control	0.55	0.05	328

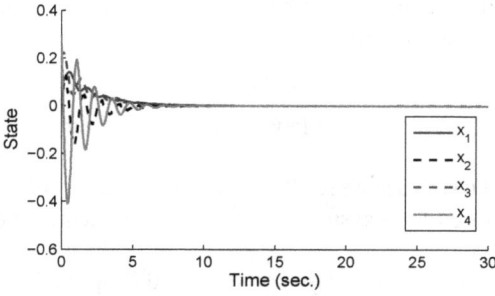

(a) Convergence of states using event-triggered control.

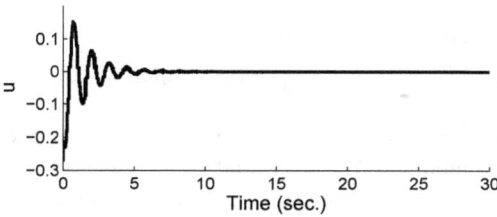

(b) Convergence of event-triggered control input $u(t_k)$.

Figure 6.5 Results of event-triggered control with system nonlinearity

6.4 ROBUST EVENT-TRIGGERED CONTROL OF BATCH REACTOR

In this section, the event-triggered control of an uncertain batch reactor system has been proposed. The unstable batch reactor, a coupled two input two output system, is a benchmark example of NCS. The contribution of this section is divided into two parts. Firstly, a robust control algorithm is proposed for a continuous-time linear system with norm bounded mismatched uncertainty. Then, an event-triggered based control algorithm is proposed. Consider an uncertain linear system as

$$\dot{x} = (A + \Delta A)x + Bu(t) \tag{6.32}$$

where $x \in \mathbb{R}^n$ and $u \in \mathbb{R}^m$ are system state and input vector, respectively. Let an unknown matrix ΔA represents the uncertainty of state-transition matrix A. In [14, 51], the norm bounded uncertainty ΔA^1 has been represented as

$$\Delta A = DF(t)E \tag{6.33}$$

---

[1]In Chapter 2, the mismatched uncertainty ΔA has been expressed as $\Delta A = BB^+(A(p) - A(p_0)) + (I - BB^+)(A(p) - A(p_0))$ where $BB^+(A(p) - A(p_0))$ and $(I - BB^+)(A(p) - A(p_0))$ are matched and mismatched components, respectively.

where D and E are known matrices with appropriate dimension. The unknown matrix $F(t)$ holds the following property

$$F(t)^T F(t) \leq I. \tag{6.34}$$

To stabilize (6.32), a robust control law is necessary which can tolerate the uncertainty ΔA.

Control objective: Design a state feedback control law $u(t) = Kx(t)$ for (6.32), such that the uncertain closed loop system (6.32) is asymptotically stable. Here K is the controller gain matrix.

To derive the robust control law for (6.32), the optimal control approach is adopted. The virtual nominal system and cost-functional for (6.32) are defined to formulate an optimal control problem:

$$\dot{x} = Ax(t) + Bu(t) + Dv(t) \tag{6.35}$$

$$J = \int_0^\infty (x^T Qx + \varepsilon^{-1} x^T E^T Ex + u^T R_1 u + v^T R_2 v) dt \tag{6.36}$$

where ε is a positive scalar and the matrices $Q \geq 0$, $R_1 > 0$, and $R_2 > 0$. The virtual system (6.35) has two inputs $u(t) = Kx(t)$ and $v(t) = Lx(t)$ where K and L are gain matrices. Here inputs $u(t)$ and $v(t)$ are stabilizing and virtual control signals, respectively. The robust control law for (6.32) is designed by solving an optimal control problem for (6.35) and (6.36). The Theorem 16 stated below is introduced to state the results.

Theorem 16 *Suppose there exist a scalar $\varepsilon > 0$ and a positive definite solution P of the Riccati equation*

$$A^T P + PA - PBR_1^{-1} B^T P - PDR_2^{-1} D^T P + Q + \varepsilon^{-1} E^T E = 0 \tag{6.37}$$

with

$$(\varepsilon^{-1} I - D^T PD) > 0. \tag{6.38}$$

The controller gain $K = -R_1^{-1} B^T P$ is the robust solution for (6.32) if it satisfies the following matrix inequality

$$Q_1 = (Q + K^T R_1 K + M^T D^T PDM - L^T R_2 L - \varepsilon^{-1} M^T M) > 0 \tag{6.39}$$

where the matrices $M = (\varepsilon^{-1} I - D^T PD)^{-1} D^T P$ and $L = -R_2^{-1} D^T P$.

    Proof *Let $V(x) = x^T Px$ is the Lyapunov function for (6.42). Then the $\dot{V}(x)$ along the direction of state trajectory of (6.42) is*

$$\dot{V}(x) = x^T (A^T P + PA + E^T F^T D^T P + PDFE - 2PBR_1^{-1} B^T P)x. \tag{6.40}$$

For further simplification of (6.40), Lemma 7 and Riccati equation (6.37) are used. It helps to simplify (6.40) as follows

$$\dot{V}(x) \leq -x^T Qx - x^T K^T R_1 Kx + x^T L^T R_2 Lx + x^T PD(\varepsilon^{-1} I - D^T PD)^{-1} D^T Px.$$

Now applying Lemma 8, the $\dot{V}(x)$ is simplified as

$$\dot{V}(x) \leq -\lambda_{min}(Q_1)\|x\|^2 \qquad (6.41)$$

where matrix Q_1 is defined in (6.39). The equation (6.41) ensures the stability of (6.32).

To stabilize a mismatched system with the aperiodic feedback information, an event-based control technique is proposed. For an event-triggered control input $u(t_k) = Kx(t_k)$, the system (6.32) is modeled as

$$\dot{x} \;\; = \;\; (A+DF(t)E)x(t)+BKx(t)+BKe(t) \qquad (6.42)$$

where $e(t)(= x(t_k) - x(t))$ is the measurement error. Using gain matrices K and L [from Theorem 16], the sufficient conditions are derived under communication constraint. A theorem is introduced next to state the results.

Theorem 17 *Let $P > 0$ be the solution of the Riccati equation (6.37) for a scalar $\varepsilon > 0$ and satisfy the inequalities (6.38), (6.34), and (6.39) where the controller gain matrices are computed as $K = -R_1^{-1}B^T P$ and $L = -R_2^{-1}D^T P$. The event-triggered control input $u(t_k) = Kx(t_k)$ ensures the ISS of (6.42) if the control input is actuated based on the following triggering sequence*

$$t_0 = 0, t_{k+1} = inf\left\{\; t \in \mathbb{R} \,|\, t \geq t_k \wedge (\mu\,\|x(t)\|^2 - \|e(k)\|^2 \leq 0)\;\right\}. \qquad (6.43)$$

The design parameter μ is defined as

$$\mu = \frac{\sigma \lambda_{min}^2(Q_1)}{4\,\|PBK\|^2}, \; \forall\, \sigma \in (0,1). \qquad (6.44)$$

To prove the above theorem, some additional results are stated in the form of lemmas. The proofs of these lemmas are similar to the proof of Lemma 4 and 5, respectively and hence details are omitted here.

Lemma 7 *Let the matrices D, E, and $F(t)$ are with appropriate dimensions where $F(t)^T F(t) \leq I$. Suppose there exist a scalar $\varepsilon > 0$ and a positive-definite solution P of (6.37) which satisfy the inequality*

$$(\varepsilon^{-1}I - D^T PD) > 0, \qquad (6.45)$$

then

$$E^T F^T D^T P + PDFE + E^T F^T D^T PDFE$$
$$\leq PD(\varepsilon^{-1}I - D^T PD)^{-1}D^T P + \varepsilon^{-1}E^T E. \qquad (6.46)$$

Lemma 8 *Let $P > 0$ be the solution of (6.37) which satisfies (6.45). For a selection of a matrix $M = (\varepsilon^{-1}I - D^T PD)^{-1} D^T P$, the following equality is obtained*

$$PD(\varepsilon^{-1}I - D^T PD)^{-1} D^T P = \varepsilon^{-1}M^T M - M^T D^T PDM. \qquad (6.47)$$

Proof of Theorem 17 *Let $V(x) = x^T Px$ is a Lyapunov function for (6.42). The $\dot{V}(x)$ along the solution of (6.42) is*

$$\dot{V}(x) = x^T (A^T P + PA + E^T F^T D^T P + PDFE - 2PBR_1^{-1} B^T P)x + 2x^T PBKe. \quad (6.48)$$

For further simplification, using Lemma 7 and Riccati equation (6.37) following inequality is obtained

$$\dot{V}(x) \leq -x^T Qx - x^T K^T R_1 Kx + x^T L^T R_2 Lx + x^T PD(\varepsilon^{-1}I - D^T PD)^{-1} D^T Px + 2x^T PBKe.$$

Now applying Lemma 8, the $\dot{V}(x)$ is simplified as

$$\dot{V}(x) \leq -\frac{\lambda_{min}(Q_1)}{2} \|x\|^2 + \frac{2}{\lambda_{min}(Q_1)} \|PBK\|^2 \|e\|^2 \qquad (6.49)$$

where matrix Q_1 is defined in (6.39). The equation (6.49) ensures the ISS of (6.42) with respect to measurement error $e(t)$. The event-triggering condition (6.43) is also derived from (6.49) using the results stated in Definition 8.

6.4.1 SIMULATION RESULTS

The control algorithm described in this section is applied to control a batch reactor system. In [130, 95], the continuous-time linearized model of a batch reactor system have been mentioned, where system matrices $A =$

$$\begin{bmatrix} 1.38 & -0.20 & 6.71 & -5.67 \\ -0.58 & -4.29 & 0 & 0.675 \\ 1.06 & 4.27 & -6.65 & 5.89 \\ 0.04 & 4.27 & 1.34 & 2.10 \end{bmatrix} \text{ and } B = \begin{bmatrix} 0 & 0 \\ 5.67 & 0 \\ 1.13 & -3.14 \\ 1.13 & 0 \end{bmatrix}. \text{ To generate the}$$

disturbance ΔA from (6.33), the known matrices $D = \begin{bmatrix} 0.1 & 0.1 & 0.2 & 0.2 \end{bmatrix}^T$ and $E = \begin{bmatrix} 0 & 0.01 & 0.2 & 0.3 \end{bmatrix}$ are considered. The unknown part of uncertainty (6.33) satisfies the condition $-1 < F(t) < 1$. For the purpose of simulation, $F(t) = 0.5sin(10t)$ is considered. The scalar $\varepsilon = 0.1$ and design matrices $Q = I$, $R_1 = 0.5I$, and $R_2 = 0.8I$ are selected. The gain matrices $K = -R_1^{-1} B^T P$ and $L = R_2^{-1} D^T P$ are

computed for the matrix $P = \begin{bmatrix} 0.7966 & -0.0528 & 0.4197 & -0.6448 \\ -0.0528 & 0.1228 & -0.0006 & 0.2061 \\ 0.4197 & -0.0006 & 0.3215 & -0.2625 \\ -0.6448 & 0.2061 & -0.2625 & 1.2240 \end{bmatrix}.$

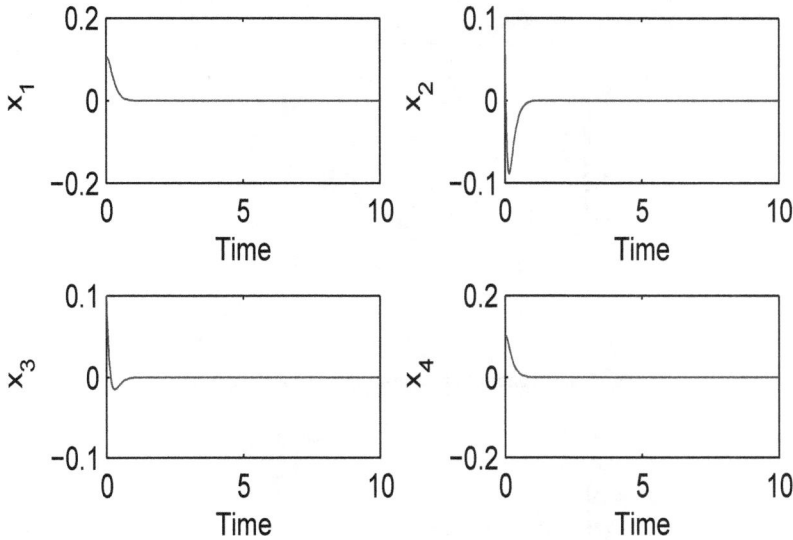

Figure 6.6 Convergence of system states for $F(t) = 0.5\sin(10t)$ with event-triggered control input

The simulation is carried out in Matlab for 10 seconds. The initial state vector is selected as $\begin{bmatrix} 0.1 & 0.1 & 0.1 & 0.1 \end{bmatrix}^T$. To realize the event-triggering law (6.43), the numerical value of parameter $\mu = 0.001$ is computed from (6.44). Figure 6.6 shows that the proposed event-triggered control input ensures the convergence of system states. The variation of control inputs are shown in Fig. 6.7. Table 6.3 is introduced to prove the efficacy of proposed control algorithm over the conventional continuous feedback approach. From Table 6.3, it is observed that for the given run-time, the total number of actuation in event-triggered control is comparatively lower than the conventional continuous approach.

Table 6.3

Comparative Results of Event-triggered and Continuous Control

Control mechanism	$\tau_{max}(sec.)$	$\tau_{min}(sec.)$	u_{total}
Continuous control	0.01	0.01	1000
Event-triggered control	0.12	0.01	634

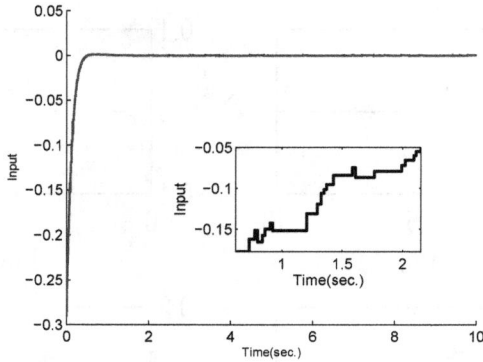

(a) Convergence of of event-triggered control input u_1.

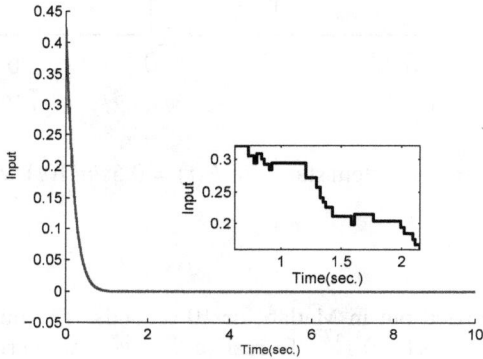

(b) Convergence of event-triggered control input u_2.

Figure 6.7 Variation of event-triggered control inputs

6.5 SUMMARY

In this chapter, the proposed optimal control framework to solve a robust control problem with limited information is applied to different class of systems. In Section 6.2 and 6.3, a manipulator dynamics and a Lipschitz nonlinear system are considered for design and analysis. The nonlinearity in system dynamics and parameter variations are considered as a source of uncertainty. To stabilize such system, the linear part of system model are considered to design the controller gain. Then, a linear robust control law is derived to stabilize such systems with limited feedback information. The triggering condition and stability results are derived based on the ISS theory and illustrates the proposed control algorithm through numerical simulation. The numerical results show the efficacy of the proposed control law over the conventional continuous feedback control approach. In the last section, robust event-triggered control law is derived using norm bounded mismatched uncertainty. Similar to Chapter 2, feedback law is computed by solving a set of matrix inequalities. Results are illustrated on a benchmark example of batch reactor system.

A Some Useful Mathematical Results

In this appendix, some useful theoretical results [61], [65], [91] are introduced. These results help to derive the proof of theorems and lemmas reported in this monograph.

MATRIX INVERSION LEMMA

Suppose A, B, C, and D are matrices with appropriate dimension with an assumption that A, D, and $(D^{-1} + CA^{-1}B)$ are invertible then there exists an identity relation as

$$(A + BDC)^{-1} = A^{-1} - A^{-1}B(D^{-1} + CA^{-1}B)^{-1}CA^{-1}. \tag{A.1}$$

SCHUR COMPLEMENT

Let $M = \left[\begin{array}{c|c} A & B \\ \hline C & D \end{array}\right] \in \mathbb{R}^{(p+q)\times(p+q)}$ is a matrix where $A \in \mathbb{R}^{p\times p}$, $B \in \mathbb{R}^{p\times q}$, $C \in \mathbb{R}^{q\times p}$, and $D \in \mathbb{R}^{q\times q}$ are given block matrices and D is invertible. Then,

Schur complement of block D of matrix M: $M/D = A - BD^{-1}C \in \mathbb{R}^{p\times p}$
Schur complement of block A of matrix M: $M/D = D - CA^{-1}B \in \mathbb{R}^{q\times q}$

Property: Let $M = \left[\begin{array}{c|c} A & B \\ \hline B^T & C \end{array}\right]$ is a symmetric matrix. The matrix M is positive definite (that means $M > 0$) if

$A > 0$ and $M/A = C - B^T A^{-1}B > 0$
$C > 0$ and $M/C = A - B^T C^{-1}B > 0$

MATRIX DIFFERENTIATION

Let P and x represent a symmetric matrix and vector, respectively. The derivative of quadratic forms are

$$\frac{\partial}{\partial x}(x^T P x) = 2Px, \tag{A.2}$$

$$\frac{\partial^2}{\partial x^2}(x^T P x) = 2P. \tag{A.3}$$

DOI: 10.1201/9781003229698-A

MATRIX INEQUALITY

For any matrices A_1, A_2, and a scalar $\varepsilon > 0$, the following inequality always holds

$$2A_1A_2 \leq \varepsilon A_1^T A_1 + \frac{1}{\varepsilon} A_2^T A_2. \tag{A.4}$$

LYAPUNOV EQUATION [47]

Let A, Q are two matrices where Q is positive definite and A is Hurwitz. Then, there exists an unique solution $P > 0$, which satisfies the following equality

$$PA + A^T P + Q = 0. \tag{A.5}$$

The solution P of (A.5) is rewritten as

$$P = \int_0^\infty e^{A^T t} Q e^{At} dt. \tag{A.6}$$

The equality (A.5) is called Lyapunov equation.

NUMERICAL SOLUTION OF RICCATI EQUATION

The discrete-time Riccati equation

$$A^T \left\{ P^{-1} + BR_1^{-1} B^T + \alpha^2 (I - BB^+) SR_2^{-1} S^T (I - BB^+)^T \right\}^{-1}$$
$$A - P + Q + 2F + 2H + \beta^2 I = 0 \tag{A.7}$$

can be written as

$$A^T \{ P^{-1} + \tilde{B}\tilde{R}^{-1} \tilde{B}^T \}^{-1} A - P + \tilde{Q} = 0 \tag{A.8}$$

where matrices $\tilde{Q}, \tilde{R}$, and $\tilde{B}$ are defined as $\tilde{Q} = Q + F + \beta^2 I$, $\tilde{B} = \begin{bmatrix} B & \alpha(I - BB^+)S \end{bmatrix}$, and $\tilde{R} = \begin{bmatrix} R_1^{-1} & 0 \\ 0 & R_2^{-1} \end{bmatrix}$, respectively. Similarly (3.29) can be reduced to in the form of (A.8), where $\tilde{Q} = Q + F + H + \beta^2 I$. To compute the numerical solution of (3.10) and (2.29), the functions '*dare*' and '*care*' are used in Matlab.

COMPARISON LEMMA

Consider a state equation as

$$\dot{x} = f(t, x) \tag{A.9}$$

where f is a nonlinear function of state x. To compute the bound of $x(t)$ without solving the equation (A.9), the comparison lemma is used. Generally in comparison lemma, for a selection of differentiable scalar function $v(x)$, the solution of differential inequality $\dot{v} \leq f(t, v)$ is compared with the solution of differential equality $\dot{u} = f(t, u)$. The primary concepts are briefly discussed in the following [71]:

Lemma 9 *Let the scalar differential equation as*

$$\dot{u} = f(t,u), \; u(t_0) = u_0 \tag{A.10}$$

where the function $f(t,u)$ hold the following properties:

> *i) $f(t,u)$ is continuous in t, $\forall t > 0$*

> *ii) $f(t,u)$ locally Lipschitz on u, $\forall t > 0$ and $u \in J \in \mathbb{R}$.*

Consider $[t_0, T)$ is the maximum time-interval for existence of the solution $u(t) \in J$ where T can be ∞. Let $v(t)$ is a scalar differential function whose upper right-hand derivative $D^+ v(t)$ satisfies

$$D^+ v(t) \leq f(t, v(t)), \; v(t_0) \leq u_0 \text{ and } v(t) \in J, \; \forall t \in [t_0, T). \tag{A.11}$$

Then, the inequality $v(t) \leq u(t)$ holds $\forall t \in [t_0, T)$.

B Proof of Lemmas

Proof of Lemma 1 *Following steps are adopted to design $u^*(t)$.*

Step 1. *Suppose $J^*(x,t)$ is the minimal cost of (1.9). From optimal control theory, the optimal input $u^*(t)$ satisfies the Hamilton-Jacobi-Bellman (HJB) equation as mentioned below*

$$-\frac{\partial J^*}{\partial t} = min_{u(t) \in \mathbb{R}^m} \left\{ x^T Qx + u^T Ru + \left(\frac{\partial J^*}{\partial x} \right)^T f(x,u) \right\}. \tag{B.1}$$

Step 2. *For a system (1.8) and cost-functional (1.9), the Hamiltonian H is*

$$\left\{ x^T Qx + u^T Ru + \left(\frac{\partial J^*}{\partial x} \right)^T f(x,u) \right\} = 0. \tag{B.2}$$

Now, the optimal input minimizes the Hamiltonian H. That means

$$\frac{\partial H}{\partial u} = 0, \tag{B.3}$$

$$2Ru^* + 2B^T \left(\frac{\partial J^*}{\partial x} \right) = 0. \tag{B.4}$$

Step 3. *Equation (B.3) helps to derive the optimal input as*

$$u^*(t) = -R^{-1}B^T \left(\frac{\partial J^*}{\partial x} \right). \tag{B.5}$$

The above mentioned steps 1 to 3 are also applicable to design the optimal input $u^*(t)$ for the linear system

$$\dot{x}(t) = Ax(t) + Bu(t). \tag{B.6}$$

For a cost functional (1.9), the optimal input u^* for (B.6) is

$$u^*(t) = -R^{-1}B^T Px(t) \tag{B.7}$$

where the positive-definite matrix $P \in \mathbb{R}^{n \times n}$ is the solution of following Riccati equation

$$PA + A^T P - PBR^{-1}B^T P + Q = 0. \tag{B.8}$$

DOI: 10.1201/9781003229698-B

Proof of Lemma 2 *To achieve $u^*(x_k)$, following steps are introduced:*

Step 1 *Discrete-time HJB: The equation (1.14) is rewritten as*

$$J(x_k) = x_k^T Q x_k + u_k^T R u_k + \sum_{n=k+1}^{\infty} \left(x_n^T Q x_n + u_n^T R u_n \right) \tag{B.9}$$

$$= x_k^T Q x_k + u_k^T R u_k + J(x_{k+1}). \tag{B.10}$$

Now from principle of optimality, for the infinite-time optimal control problem, the minimum value function $J^*(x_k, u_k)$ satisfies the following DT-HJB equation

$$J^*(x_k) = min_{u_k}\{x_k^T Q x_k + u_k^T R u_k + J^*(x_{k+1})\}. \tag{B.11}$$

Step 2 *Optimal input: From optimal control theory, it is known that, the optimal input $u^*(x_k)$ minimizes the following Hamiltonian*

$$H(x_k, u_k) = x_k^T Q x_k + u_k^T R u_k + J(x_k)^* - J(x_{k+1})^* \tag{B.12}$$

that means,

$$min_{u_k}\left\{ H(x_k, u_k^*) \right\} = 0. \tag{B.13}$$

Now using the gradient of (B.12), the optimal input u^* is achieved as

$$u_k^* = -\frac{1}{2}R^{-1}g^T(x_k)\frac{\partial J^*(x_{k+1})}{\partial x_{k+1}}. \tag{B.14}$$

Substituting (B.14) in (B.11), the Dt-HJB equation is rewritten as

$$x_k^T Q x_k + \frac{1}{4}\frac{\partial J^*(x_{k+1})}{\partial x_{k+1}}g(x_k)R^{-1}g(x_k)^T\frac{\partial J^*(x_{k+1})}{\partial x_{k+1}} + J(x_{k+1}) - J(x_k) = 0. \tag{B.15}$$

Using cost-functional (1.14), the above steps can be adopted to compute the optimal input for the linear system

$$x_{k+1} = Ax_k + Bu_k. \tag{B.16}$$

In that case, the optimal input u_k^* is computed as

$$u_k^* = -\frac{1}{2}R^{-1}g(x_k)Px_k \tag{B.17}$$

where matrix $P > 0$ is a solution of following Riccati equation

$$A^T\{P^{-1} + BR^{-1}B^T\}^{-1}A - P + Q = 0. \tag{B.18}$$

C Discretization Method

The conversion procedure of a continuous-time system to a discrete-time system is described in this appendix. For this purpose, the discretization method outlined in [23] [125] is adopted. Let there exists a continuous-time system as

$$\dot{x}(t) = Ax(t) + Bu \qquad (C.1)$$

$$y(t) = Cx(t) + Du. \qquad (C.2)$$

Consider the time-instants $t_2 < t_1$, the equation (C.1) can be integrated as

$$x(t_2) = e^{(t_2 - t_1)A}x(t_1) + \int_{t_1}^{t_2} e^{t_2 - \tau}ABu(\tau)d\tau. \qquad (C.3)$$

For digital implementation of control input $u(t)$, the time t is expressed as $t = kT$ where k and T are the discrete-time index and sampling period, respectively. Let the time-instants $t_1 = kT$ and $t_2 = (k+1)T$ for which equation (C.3) is written as

$$x[(k+1)T] = e^{TA}x(t_1) + \int_{t_1}^{t_2} e^{[(k+1)T - \tau]A}Bu(\tau)d\tau. \qquad (C.4)$$

The input $u(\tau)$ remains constant over the kth sampling period. That means

$$u(t) = u(k), \quad \forall \, kT \le t \le (k+1)T. \qquad (C.5)$$

Hence, using (C.5), (C.4) can be written as

$$x[(k+1)T] = e^{AT}x(kT) + \left[\int_{kT}^{(k+1)T} e^{[(k+1)T - \tau]A}Bd\tau \right]u(k). \qquad (C.6)$$

Suppose $x[k] = x(kT)$ represents the state vector $x(t)$ at $t = kT$ sampling instant. Then the equation (C.6) is simplified as

$$x(k+1) = \Phi x(k) + \Gamma u(k) \qquad (C.7)$$

where $\Phi = e^{AT}$ and $\Gamma = \left[\int_{kT}^{(k+1)T} e^{[(k+1)T - \tau]A}Bd\tau \right]$.

Numerical computation: For a given state space description A, B and sampling time T, the ZOH equivalent model (C.7) can be computed by finding a matrix exponential of MT. The matrix $M = \begin{bmatrix} A & B \\ \hline 0 & 0 \end{bmatrix}$ is a partitioned matrix which holds the following relation [125]:

$$e^{MT} = \begin{bmatrix} \Phi & \Gamma \\ \hline 0 & 1 \end{bmatrix}. \qquad (C.8)$$

DOI: 10.1201/9781003229698-C

References

1. Mahmoud Abdelrahim, Romain Postoyan, Jamal Daafouz, and Dragan Nešić. Stabilization of nonlinear systems using event-triggered output feedback controllers. *IEEE Transactions on Automatic Control*, 61(9):2682–2687, 2016.

2. Mahmoud Abdelrahim, Romain Postoyan, Jamal Daafouz, and Dragan Nešić. Robust event-triggered output feedback controllers for nonlinear systems. *Automatica*, 75:96–108, 2017.

3. Dipak M Adhyaru, IN Kar, and M Gopal. Bounded robust control of nonlinear systems using neural network–based hjb solution. *Neural Computing and Applications*, 20(1):91–103, 2011.

4. DM Adhyaru, IN Kar, and M Gopal. Fixed final time optimal control approach for bounded robust controller design using Hamilton–Jacobi–Bellman solution. *IET control theory & applications*, 3(9):1183–1195, 2009.

5. Asma Al-Tamimi, Frank L Lewis, and Murad Abu-Khalaf. Discrete-time nonlinear hjb solution using approximate dynamic programming: Convergence proof. *IEEE Transactions on Systems, Man, and Cybernetics, Part B (Cybernetics)*, 38(4):943–949, 2008.

6. R Alur, K-E Arzen, John Baillieul, TA Henzinger, Dimitrios Hristu-Varsakelis, and William S Levine. *Handbook of networked and embedded control systems*. Springer Science & Business Media, 2007.

7. Brian DO Anderson and John B Moore. *Optimal control: linear quadratic methods*. Courier Corporation, 2007.

8. Adolfo Anta and Paulo Tabuada. On the benefits of relaxing the periodicity assumption for networked control systems over can. In *Real-Time Systems Symposium, 2009, RTSS 2009. 30th IEEE*, pages 3–12. IEEE, 2009.

9. Adolfo Anta and Paulo Tabuada. To sample or not to sample: Self-triggered control for nonlinear systems. *IEEE Transactions on Automatic Control*, 55(9):2030–2042, 2010.

10. José Araújo, Manuel Mazo, Adolfo Anta, Paulo Tabuada, and Karl H Johansson. System architectures, protocols and algorithms for aperiodic wireless control systems. *IEEE Transactions on Industrial Informatics*, 10(1):175–184, 2014.

11. Karl-Erik Arzén. A simple event-based pid controller. In *Proc. 14th IFAC World Congress*, volume 18, pages 423–428, 1999.

12. Karl J Aström. Event based control. In *Analysis and design of nonlinear control systems*, pages 127–147. Springer, 2008.

13. Karl Johan ström and Bo Bernhardsson. Comparison of riemann and lebesque sampling for first order stochastic systems. In *Proceedings of the 41st IEEE Conference on Decision and Control, 2002*, volume 2, pages 2011–2016. IEEE, 2002.

14. B Barmish. Stabilization of uncertain systems via linear control. *IEEE Trans. Autom. Control*, 28(8):848–850, 1983.

15. B Ross Barmish. Necessary and sufficient conditions for quadratic stabilizability of an uncertain system. *Journal of Optimization theory and applications*, 46(4):399–408, 1985.

16. BR Barmish, M Corless, and G Leitmann. A new class of stabilizing controllers for uncertain dynamical systems. *SIAM Journal on Control and Optimization*, 21(2):246–255, 1983.

17. Randal W Beard. *Improving the closed-loop performance of nonlinear systems*. PhD thesis, Citeseer, 1995.

18. Abhisek Behera and Bijnan Bandyopadhyay. Robust sliding mode control: an event-triggering approach. *IEEE Transactions on Circuits and Systems II: Express Briefs*, 2016.

19. Dimitri P Bertsekas and John N Tsitsiklis. Neuro-dynamic programming: an overview. In *Decision and Control, 1995., Proceedings of the 34th IEEE Conference on*, volume 1, pages 560–564. IEEE, 1995.

20. Mahendra Bhadu, Niladri Sekhar Tripathy, Indra Narayan Kar, and Nilanjan Senroy. Event-triggered communication in wide-area damping control: a limited output feedback-based approach. *IET Generation, Transmission & Distribution*, 10(16):4094–4104, 2016.

21. FD Brunne, WPMH Heemels, and F Allgower. Dynamic thresholds in robust event-triggered control for discrete-time linear systems. In *Control Conference (ECC), 2016 European*, pages 923–988. IEEE, 2016.

22. Fu-Chuang Chen and Chen-Chung Liu. Adaptively controlling nonlinear continuous-time systems using multilayer neural networks. *IEEE Transactions on Automatic Control*, 39(6):1306–1310, 1994.

23. Tongwen Chen and Bruce A Francis. *Optimal sampled-data control systems*. Springer Science & Business Media, 2012.

24. YH Chen. Design of robust controllers for uncertain dynamical systems. *IEEE Transactions on Automatic Control*, 33(5):487–491, 1988.

25. Zheng Chen and Sarangapani Jagannathan. Generalized Hamilton–Jacobi–Bellman formulation-based neural network control of affine nonlinear discrete-time systems. *IEEE Transactions on Neural Networks*, 19(1):90–106, 2008.

26. Tayfun Cimen. State-dependent Riccati equation (SDRE) control: A survey. *IFAC Proceedings Volumes*, 41(2):3761–3775, 2008.

27. Marieke BG Cloosterman, Laurentiu Hetel, Nathan Van de Wouw, WPMH Heemels, Jamal Daafouz, and Henk Nijmeijer. Controller synthesis for networked control systems. *Automatica*, 46(10):1584–1594, 2010.

28. James R Cloutier. State-dependent Riccati equation techniques: an overview. In *American Control Conference, 1997. Proceedings of the 1997*, volume 2, pages 932–936. IEEE, 1997.

29. John J Craig. *Introduction to robotics: mechanics and control*, volume 3. Pearson Prentice Hall, India, 2005.

30. Michele Cucuzzella, Gian Paolo Incremona, and Antonella Ferrara. Event-triggered sliding mode control algorithms for a class of uncertain nonlinear systems: experimental assessment. In *American Control Conference (ACC), 2016*, pages 6549–6554. IEEE, 2016.

31. Dragan B Dačić and Dragan Nešić. Quadratic stabilization of linear networked control systems via simultaneous protocol and controller design. *Automatica*, 43(7):1145–1155, 2007.

32. Dimos V Dimarogonas, Emilio Frazzoli, and Karl H Johansson. Distributed event-triggered control for multi-agent systems. *IEEE Transactions on Automatic Control*, 57(5):1291–1297, 2012.

33. Derui Ding, Zidong Wang, Bo Shen, and Guoliang Wei. Event-triggered consensus control for discrete-time stochastic multi-agent systems: the input-to-state stability in probability. *Automatica*, 62:284–291, 2015.

34. VS Dolk, DP Borgers, and WPMH Heemels. Output-based and decentralized dynamic event-triggered control with guaranteed l_p-gain performance and zeno-freeness. *IEEE Transactions on Automatic Control*, 62(1):34–49, 2017.

35. Chuan Dong, Suiqiong Li, Mengyang Li, Qisheng He, Dacheng Xu, and Xinxin Li. Self-powered event-triggered wireless sensor network for monitoring sabotage activities. In *SENSORS, 2016 IEEE*, pages 1–3. IEEE, 2016.

36. Lu Dong, Yufei Tang, Haibo He, and Changyin Sun. An event-triggered approach for load frequency control with supplementary adp. *IEEE Transactions on Power Systems*, 32(1):581–589, 2017.

37. MCF Donkers and WPMH Heemels. Output-based event-triggered control with guaranteed-gain and improved and decentralized event-triggering. *IEEE Transactions on Automatic Control*, 57(6):1362–1376, 2012.

38. RC Dorf, M Farren, and C Phillips. Adaptive sampling frequency for sampled-data control systems. *IRE Transactions on Automatic Control*, 7(1):38–47, 1962.

39. Nicola Elia and Sanjoy K Mitter. Stabilization of linear systems with limited information. *IEEE transactions on Automatic Control*, 46(9):1384–1400, 2001.

40. Alina Eqtami, Dimos V Dimarogonas, and Kostas J Kyriakopoulos. Event-triggered control for discrete-time systems. In *Proceedings of the 2010 american control conference*, pages 4719–4724. IEEE, 2010.

41. Tomas Estrada and Panos J Antsaklis. Model-based control with intermittent feedback: Bridging the gap between continuous and instantaneous feedback. *International Journal of Control*, 83(12):2588–2605, 2010.

42. Yuan Fan, Gang Feng, Yong Wang, and Cheng Song. Distributed event-triggered control of multi-agent systems with combinational measurements. *Automatica*, 49(2):671–675, 2013.

43. F Felicioni, N Jia, F Simonot-Lion, and YQ Song. Co-design approaches for dependable networked control systems, 2010.

44. Bruce A Finlayson. *The method of weighted residuals and variational principles.* SIAM, 2013.

45. Hisaya Fujioka. Stability analysis for a class of networked/embedded control systems: A discrete-time approach. In *American Control Conference, 2008*, pages 4997–5002. IEEE, 2008.

46. Katsuhisa Furuta, Masaki Yamakita, and Seiichi Kobayashi. Swing up control of inverted pendulum. In *Industrial Electronics, Control and Instrumentation, 1991. Proceedings. IECON'91, 1991 International Conference on*, pages 2193–2198. IEEE, 1991.

47. Zoran Gajic and Muhammad Tahir Javed Qureshi. *Lyapunov matrix equation in system stability and control.* Courier Corporation, New York, 2008.

48. Yong-Feng Gao, Rui Wang, Changyun Wen, and Wei Wang. Digital event-based control for nonlinear systems without the limit of iss. *IEEE Transactions on Circuits and Systems II: Express Briefs*, 2016.

49. Eloy Garcia and Panos J Antsaklis. Optimal model-based control with limited communication. *IFAC Proceedings Volumes*, 47(3):10908–10913, 2014.

50. Eloy Garcia, Panos J Antsaklis, and Luis A Montestruque. *Model-based control of networked systems.* Springer, 2014.

51. Germain Garcia, Denis Arzelier, et al. Robust stabilization of discrete-time linear systems with norm-bounded time-varying uncertainty. *Systems & Control Letters*, 22(5):327–339, 1994.

52. Antoine Girard. Dynamic triggering mechanisms for event-triggered control. *IEEE Transactions on Automatic Control*, 60(7):1992–1997, 2015.

53. CJ Goh. On the nonlinear optimal regulator problem. *Automatica*, 29(3):751–756, 1993.

54. SC Gupta. Increasing the sampling efficiency for a control system. *IEEE Transactions on Automatic Control*, 8(3):263–264, 1963.

55. WP Maurice H Heemels, Andrew R Teel, Nathan Van de Wouw, and Dragan Nesic. Networked control systems with communication constraints: Tradeoffs between transmission intervals, delays and performance. *IEEE Transactions on Automatic control*, 55(8):1781–1796, 2010.

56. WPMH Heemels and MCF Donkers. Model-based periodic event-triggered control for linear systems. *Automatica*, 49(3):698–711, 2013.

57. WPMH Heemels, MCF Donkers, and Andrew R Teel. Periodic event-triggered control for linear systems. *IEEE Transactions on Automatic Control*, 58(4):847–861, 2013.

58. WPMH Heemels, Karl Henrik Johansson, and Paulo Tabuada. An introduction to event-triggered and self-triggered control. In *2012 IEEE 51st IEEE Conference on Decision and Control (CDC)*, pages 3270–3285. IEEE, 2012.

59. Toivo Henningsson, Erik Johannesson, and Anton Cervin. Sporadic event-based control of first-order linear stochastic systems. *Automatica*, 44(11):2890–2895, 2008.

60. Ali Heydari and SN Balakrishnan. Closed-form solution to finite-horizon suboptimal control of nonlinear systems. *International Journal of Robust and Nonlinear Control*, 25(15):2687–2704, 2015.

61. Roger A Horn and Charles R Johnson. *Matrix analysis*. Cambridge university press, 2012.

62. Shin-Chen Hu, Chen-Yu Chan, and Yen-Chen Liu. Event-triggered control for bilateral teleoperation with time delays. In *Advanced Intelligent Mechatronics (AIM), 2016 IEEE International Conference on*, pages 1634–1639. IEEE, 2016.

63. Songlin Hu and Dong Yue. L2-gain analysis of event-triggered networked control systems: a discontinuous Lyapunov functional approach. *International Journal of Robust and Nonlinear Control*, 23(11):1277–1300, 2013.

64. Songlin Hu, Dong Yue, Xiuxia Yin, Xiangpeng Xie, and Yong Ma. Adaptive event-triggered control for nonlinear discrete-time systems. *International Journal of Robust and Nonlinear Control*, 26(18):4104–4125, 2016.

65. IN Imam. The schur complement and the inverse m-matrix problem. *Linear algebra and its applications*, 62:235–240, 1984.

66. Ohran C Imer and Tamer Basar. Optimal control with limited controls. In *American Control Conference, 2006*, pages 6–pp. IEEE, 2006.

67. Gian Paolo Incremona and Antonella Ferrara. Adaptive model-based event-triggered sliding mode control. *International Journal of Adaptive Control and Signal Processing*, 2016.

68. Zhong-Ping Jiang and Yuan Wang. Input-to-state stability for discrete-time nonlinear systems. *Automatica*, 37(6):857–869, 2001.

69. Karl Henrik Johansson, Magnus Egerstedt, John Lygeros, and Shankar Sastry. On the regularization of zeno hybrid automata. *Systems & Control Letters*, 38(3):141–150, 1999.

70. EI Jury and FJ Mullin. The analysis of sampled-data control systems with a periodically time-varying sampling rate. *IRE Transactions on Automatic Control*, (1):15–21, 1959.

71. Hassan K Khalil. *Noninear Systems*. Prentice-Hall, New Jersey, 1996.

72. Donald E Kirk. *Optimal control theory: an introduction*. Courier Corporation, 2012.

73. Hyung-Woo Lee, Ki-Chan Kim, and Ju Lee. Review of maglev train technologies. *IEEE transactions on magnetics*, 42(7):1917–1925, 2006.

74. Daniel Lehmann, Jan Lunze, and Karl Henrik Johansson. Comparison between sampled-data control, deadband control and model-based event-triggered control. *IFAC Proceedings Volumes*, 45(9):7–12, 2012.

75. Michael Lemmon. Event-triggered feedback in control, estimation, and optimization. In *Networked Control Systems*, pages 293–358. Springer, 2010.

76. FW Lewis, Suresh Jagannathan, and A Yesildirak. *Neural network control of robot manipulators and non-linear systems*. CRC Press, 1998.

77. Lichun Li and Michael Lemmon. Event-triggered output feedback control of finite horizon discrete-time multi-dimensional linear processes. In *Decision and Control (CDC), 2010 49th IEEE Conference on*, pages 3221–3226. IEEE, 2010.

78. Wenling Li, Yingmin Jia, and Junping Du. Event-triggered kalman consensus filter over sensor networks. *IET Control Theory & Applications*, 10(1):103–110, 2016.

79. Feng Lin. An optimal control approach to robust control design. *International journal of control*, 73(3):177–186, 2000.

80. Feng Lin and Robert D Brandt. An optimal control approach to robust control of robot manipulators. *IEEE Transactions on Robotics and Automation*, 14(1):69–77, 1998.

81. Feng Lin and Andrzej W Olbrot. An lqr approach to robust control of linear systems with uncertain parameters. In *Decision and Control, 1996., Proceedings of the 35th IEEE Conference on*, volume 4, pages 4158–4163. IEEE, 1996.

82. Feng Lin, William Zhang, and Robert D Brandt. Robust hovering control of a PVTOL aircraft. *IEEE Transactions on Control Systems Technology*, 7(3):343–351, 1999.

83. Davide Liuzza, Dimos V Dimarogonas, Mario Di Bernardo, and Karl H Johansson. Distributed model based event-triggered control for synchronization of multi-agent systems. *Automatica*, 73:1–7, 2016.

84. Nicolas Marchand, Sylvain Durand, and Jose Fermi Guerrero Castellanos. A general formula for event-based stabilization of nonlinear systems. *IEEE Transactions on Automatic Control*, 58(5):1332–1337, 2013.

85. Pau Martí, Manel Velasco, Antonio Camacho, Enric Xavier Martín, and Josep M Fuertes. Networked sliding mode control of the double integrator system using the event-driven self-triggered approach. In *Industrial Electronics (ISIE), 2011 IEEE International Symposium on*, pages 2031–2036. IEEE, 2011.

86. Manuel Mazo and Paulo Tabuada. Decentralized event-triggered control over wireless sensor/actuator networks. *IEEE Transactions on Automatic Control*, 56(10):2456–2461, 2011.

87. Xiangyu Meng and Tongwen Chen. Optimal sampling and performance comparison of periodic and event based impulse control. *IEEE Transactions on Automatic Control*, 57(12):3252–3259, 2012.

88. Adam Molin. Optimal event-triggered control with communication constraints. *Technische Universität München, München*, 2014.

89. Adam Molin and Sandra Hirche. Adaptive event-triggered control over a shared network. In *Decision and Control (CDC), 2012 IEEE 51st Annual Conference on*, pages 6591–6596. IEEE, 2012.

90. Curtis P Mracek and James R Cloutier. Control designs for the nonlinear benchmark problem via the state-dependent Riccati equation method. *International Journal of robust and nonlinear control*, 8(4-5):401–433, 1998.

91. D Subbaram Naidu. *Optimal control systems*. CRC press, 2002.

92. J Nazarzadeh, M Razzaghi, and KY Nikravesh. Solution of the matrix Riccati equation for the linear quadratic control problems. *Mathematical and Computer Modelling*, 27(7):51–55, 1998.

93. D Nešić and AR Teel. Input-to-state stability of networked control systems. *Automatica*, 40(12):2121–2128, 2004.

94. Dragan Nesic and Andrew R Teel. A framework for stabilization of nonlinear sampled-data systems based on their approximate discrete-time models. *IEEE Transactions on automatic control*, 49(7):1103–1122, 2004.

95. Dragan Nesic and Andrew R Teel. Input-output stability properties of networked control systems. *IEEE Trans. Autom. Control*, 49(10):1650–1667, 2004.

96. Thang Nguyen and Zoran Gajic. Solving the matrix differential Riccati equation: a Lyapunov equation approach. *IEEE Transactions on Automatic Control*, 55(1):191–194, 2010.

97. Cameron Nowzari and Jorge Cortés. Self-triggered coordination of robotic networks for optimal deployment. *Automatica*, 48(6):1077–1087, 2012.

98. Paul G Otanez, James R Moyne, and Dawn M Tilbury. Using deadbands to reduce communication in networked control systems. In *American Control Conference, 2002. Proceedings of the 2002*, volume 4, pages 3015–3020. IEEE, 2002.

99. Chen Peng, Dong Yue, and Min-Rui Fei. A higher energy-efficient sampling scheme for networked control systems over IEEE 802.15. 4 wireless networks. *IEEE Transactions on Industrial Informatics*, 12(5):1766–1774, 2016.

100. Ian R Petersen. Structural stabilization of uncertain systems: Necessity of the matching condition. *SIAM journal on control and optimization*, 23(2):286–296, 1985.

101. G Phanomchoeng and R Rajamani. Observer design for lipschitz nonlinear systems using Riccati equations. In *Proc. American Control Conf.*, pages 6060–6065, 2010.

102. Nicholas J Ploplys, Paul A Kawka, and Andrew G Alleyne. Closed-loop control over wireless networks. *IEEE control systems*, 24(3):58–71, 2004.

103. Marios M Polycarpou. Stable adaptive neural control scheme for nonlinear systems. *IEEE Transactions on Automatic Control*, 41(3):447–451, 1996.

104. Romain Postoyan, Marcos Cesar Bragagnolo, Ernest Galbrun, Jamal Daafouz, Dragan Nešić, and Eugênio B Castelan. Event-triggered tracking control of unicycle mobile robots. *Automatica*, 52:302–308, 2015.

105. Anna Prach, Ozan Tekinalp, and Dennis S Bernstein. A numerical comparison of frozen-time and forward-propagating Riccati equations for stabilization of periodically time-varying systems. In *American Control Conference (ACC), 2014*, pages 5633–5638. IEEE, 2014.

106. Daniel E Quevedo, Vijay Gupta, Wann-Jiun Ma, and Serdar Yüksel. Stochastic stability of event-triggered anytime control. *IEEE Transactions on Automatic Control*, 59(12):3373–3379, 2014.

107. Chithrupa Ramesh, Henrik Sandberg, and Karl H Johansson. Design of state-based schedulers for a network of control loops. *IEEE Transactions on Automatic Control*, 58(8):1962–1975, 2013.

108. Henrik Rehbinder and Martin Sanfridson. Scheduling of a limited communication channel for optimal control. *Automatica*, 40(3):491–500, 2004.

109. Nader Sadegh. A perceptron network for functional identification and control of nonlinear systems. *IEEE Transactions on Neural Networks*, 4(6):982–988, 1993.

110. Avimanyu Sahoo, Hao Xu, and Sarangapani Jagannathan. Near optimal event-triggered control of nonlinear discrete-time systems using neurodynamic programming. *IEEE transactions on neural networks and learning systems*, 27(9):1801–1815, 2016.

111. Avimanyu Sahoo, Hao Xu, and Sarangapani Jagannathan. Neural network-based event-triggered state feedback control of nonlinear continuous-time systems. *IEEE transactions on neural networks and learning systems*, 27(3):497–509, 2016.

112. Jagannathan Sarangapani. *Neural network control of nonlinear discrete-time systems*, volume 21. CRC press, 2006.

113. John Seiffertt, Suman Sanyal, and Donald C Wunsch. Hamilton–Jacobi–Bellman equations and approximate dynamic programming on time scales. *IEEE Transactions on Systems, Man, and Cybernetics, Part B (Cybernetics)*, 38(4):918–923, 2008.

114. Bruno Sinopoli, Luca Schenato, Massimo Franceschetti, Kameshwar Poolla, Michael I Jordan, and Shankar S Sastry. Kalman filtering with intermittent observations. *IEEE transactions on Automatic Control*, 49(9):1453–1464, 2004.

115. Eduardo D Sontag. Input to state stability: Basic concepts and results. In *Nonlinear and optimal control theory*, pages 163–220. Springer, 2008.

116. Mohammad Tabbara, Dragan Nešić, and Andrew R Teel. Networked control systems: Emulation-based design. In *Networked control systems*, pages 57–94. Springer, 2008.

117. Paulo Tabuada. Event-triggered real-time scheduling of stabilizing control tasks. *IEEE Transactions on Automatic Control*, 52(9):1680–1685, 2007.

118. Pavankumar Tallapragada and Nikhil Chopra. On event triggered tracking for nonlinear systems. *IEEE Transactions on Automatic Control*, 58(9):2343–2348, 2013.

119. Haihua Tan, Shaolong Shu, and Feng Lin. An optimal control approach to robust tracking of linear systems. *International Journal of Control*, 82(3):525–540, 2009.

120. Sekhar Tatikonda and Sanjoy Mitter. Control under communication constraints. *IEEE Transactions on Automatic Control*, 49(7):1056–1068, 2004.

121. Sekhar Tatikonda, Anant Sahai, and Sanjoy Mitter. Control of lqg systems under communication constraints. In *Decision and Control, 1998. Proceedings of the 37th IEEE Conference on*, volume 1, pages 1165–1170. IEEE, 1998.

122. Ubaldo Tiberi, Carlo Fischione, Karl Henrik Johansson, and Maria Domenica Di Benedetto. Energy-efficient sampling of networked control systems over ieee 802.15. 4 wireless networks. *Automatica*, 49(3):712–724, 2013.

123. Sebastian Trimpe and Raffaello D'Andrea. Event-based state estimation with variance-based triggering. *IEEE Transactions on Automatic Control*, 59(12):3266–3281, 2014.

124. Niladri Sekhar Tripathy, IN Kar, and Kolin Paul. An event-triggered based robust control of robot manipulator. In *Control Automation Robotics & Vision (ICARCV), 2014 13th International Conference on*, pages 425–430. IEEE, 2014.

125. Richard J Vaccaro. *Digital control: a state-space approach*, volume 196. McGraw-Hill New York, 1995.

126. Kyriakos G Vamvoudakis. Event-triggered optimal adaptive control algorithm for continuous-time nonlinear systems. *IEEE/CAA Journal of Automatica Sinica*, 1(3):282–293, 2014.

127. Kyriakos G Vamvoudakis. An online actor/critic algorithm for event-triggered optimal control of continuous-time nonlinear systems. In *American Control Conference (ACC), 2014*, pages 1–6. IEEE, 2014.

128. Kyriakos G Vamvoudakis, Arman Mojoodi, and Henrique Ferraz. Event-triggered optimal tracking control of nonlinear systems. *International Journal of Robust and Nonlinear Control*, 2016.

129. Draguna Vrabie, O Pastravanu, Murad Abu-Khalaf, and Frank L Lewis. Adaptive optimal control for continuous-time linear systems based on policy iteration. *Automatica*, 45(2):477–484, 2009.

130. Gregory C Walsh and Hong Ye. Scheduling of networked control systems. *IEEE Control Syst.*, 21(1):57–65, 2001.

131. Ding Wang, Derong Liu, and Hongliang Li. Policy iteration algorithm for online design of robust control for a class of continuous-time nonlinear systems. *IEEE Transactions on Automation Science and Engineering*, 11(2):627–632, 2014.

132. Ding Wang, Derong Liu, Hongliang Li, Biao Luo, and Hongwen Ma. An approximate optimal control approach for robust stabilization of a class of discrete-time nonlinear systems with uncertainties. *IEEE Transactions on Systems, Man, and Cybernetics: Systems*, 46(5):713–717, 2016.

133. Ding Wang, Derong Liu, Qichao Zhang, and Dongbin Zhao. Data-based adaptive critic designs for nonlinear robust optimal control with uncertain dynamics. *IEEE Transactions on Systems, Man, and Cybernetics: Systems*, 46(11):1544–1555, 2016.

134. Xiaofeng Wang and Michael D Lemmon. Self-triggered feedback control systems with finite-gain l_2 stability. *IEEE transactions on automatic control*, 54(3):452–467, 2009.

135. Xiaofeng Wang and Michael D Lemmon. Event-triggering in distributed networked control systems. *IEEE Transactions on Automatic Control*, 56(3):586–601, 2011.

136. Christopher John Cornish Hellaby Watkins. *Learning from delayed rewards*. PhD thesis, University of Cambridge England, 1989.

137. Sean Weerakkody, Yilin Mo, Bruno Sinopoli, Duo Han, and Ling Shi. Multi-sensor scheduling for state estimation with event-based, stochastic triggers. *IEEE Transactions on Automatic Control*, 61(9):2695–2701, 2016.

138. Shiping Wen, Xinghuo Yu, Zhigang Zeng, and Jinjian Wang. Event-triggering load frequency control for multiarea power systems with communication delays. *IEEE Transactions on Industrial Electronics*, 63(2):1308–1317, 2016.

139. Norbert Wiener. Nonlinear problems in random theory. *Nonlinear Problems in Random Theory, by Norbert Wiener, pp. 142. ISBN 0-262-73012-X. Cambridge, Massachusetts, USA: The MIT Press, August 1966.(Paper)*, page 142, 1966.

140. Meng Xia, Vijay Gupta, and Panos J Antsaklis. Networked state estimation over a shared communication medium. In *2013 American Control Conference*, pages 4128–4133. IEEE, 2013.

141. Lantao Xing, Changyun Wen, Zhitao Liu, Hongye Su, and Jianping Cai. Event-triggered adaptive control for a class of uncertain nonlinear systems. *IEEE Transactions on Automatic Control*, 2016.

142. Hao Xu, Avimanyu Sahoo, and Sarangapani Jagannathan. Stochastic adaptive event-triggered control and network scheduling protocol co-design for distributed networked systems. *IET Control Theory & Applications*, 8(18):2253–2265, 2014.

143. Wenying Xu, Daniel WC Ho, Lulu Li, and Jinde Cao. Event-triggered schemes on leader-following consensus of general linear multiagent systems under different topologies. *IEEE transactions on cybernetics*, 47(1):212–223, 2017.

144. Jun Yang, Shihua Li, and Xinghuo Yu. Sliding-mode control for systems with mismatched uncertainties via a disturbance observer. *IEEE Transactions on Industrial Electronics*, 60(1):160–169, 2013.

145. Jianhua Zhang, Yamin Kuai, Mifeng Ren, Zhongli Luo, and Mingming Lin. Event-triggered distributed filtering for non-Gaussian systems over wireless sensor networks using survival information potential criterion. *IET Control Theory & Applications*, 10(13):1524–1530, 2016.

146. Lei Zhang and D Hristu-Varsakelis. Lqg control under limited communication. In *Decision and Control, 2005 and 2005 European Control Conference. CDC-ECC'05. 44th IEEE Conference on*, pages 185–190. IEEE, 2005.

147. Wei Zhang, Michael S Branicky, and Stephen M Phillips. Stability of networked control systems. *IEEE Control Systems*, 21(1):84–99, 2001.

148. Yan Zheng, Georigi M Dimirovski, Yuanwei Jing, and Muyi Yang. Discrete-time sliding mode control of nonlinear systems. In *American Control Conference, 2007. ACC'07*, pages 3825–3830. IEEE, 2007.

Index